FACING FORWARD

THE HISTORY OF WAUKESHA ENGINE

BY MICHAEL J. GOC

PREVIOUS PAGE,
Engine builders, 1910s.

ISBN 0-938627-69-4

FACING FORWARD
The History of Waukesha Engine

Publishing Services:
New Past Press Inc., Friendship, WI
Written and Edited by Michael J. Goc
Cover & page design by Jay Jocham, Big Flats/Hancock, WI
Business Management & Marketing by Carol Huber Podoll

Library of Congress
Cataloging-in-Publication Data

Goc, Michael J.
Facing forward : the history of Waukesha Engine 1906-2006 / Michael J. Goc.
p. cm.
ISBN 0-938627-69-4 (alk. paper)
1. Waukesha Engine Division, Dresser Industrial Products--History.
2. Internal combustion engines. I. Title.
TJ130.W34G63 2006
621.43'40973--dc22
2006034219

ACKNOWLEDGMENTS

Waukesha Engine is an unusual company; unusual in that it has survived for one hundred years; even more unusual in that it continues to manufacture the same product it did a century ago. Of course, in nearly every respect, the engines of 2006 are light years away from the motors of 1906, but they are still engines. Should one of the company's founders, or a pioneer shop worker, see Waukesha's latest model, he would be awed by its power, efficiency and technological sophistication, but he would still recognize it as an engine.

Waukesha Engine is unusual for its historical continuity and most unusual because it recognizes its past and has taken steps to preserve it. Records of the old Motor Company's first days and of all the decades in between that could have been destroyed, have been preserved. They constitute a remarkable archive of an American manufacturing company in the twentieth century and an invaluable historical resource.

The corporate commitment to preserve Waukesha's history is matched by the personal commitment of the retirees who have organized the Waukesha Engine Historical Society. Not only do these volunteers work with the company to preserve its records, they actively collect the stuff of history--memoirs, photos, art, correspondence, advertising, artifacts, engines--wherever they may be found. They donate their time to enlarge and document the story of the company that has played a major role in their own lives.

Without the company's commitment to preserve its past and the historical society's efforts this book could not have been written. For example, with only a few exceptions, every historical photo in this book came from the collections of the Waukesha Engine Historical Society. Most of them were the work of the superb Waukesha photographic studio of Warren O'Brien.

Historical society members provided research assistance and candid editorial opinions on this book. The input of retired Waukesha personnel, Chet Lee, Robert Harris, Ron Long, Cliff Borgstrom and Rocky Schaefer helped shape and improve the work. Help also came from active and former staff: Cary Peck, Sheila Gailloretto, Robert Stachowicz, Steve Lind, Chuck Lawton, Lloyd Beauchamp, Paul Cannestra, Thomas Laird and Patrick Murray.

A special word of thanks goes to Nancy Deptolla, who shepherded the project from start to finish.

It has been a privilege to tell their story.

Michael J. Goc
October, 2006

CONTENTS

OPENING DECADES 7

What's In A Name 10
A NEW COMPANY 17
The Golden Age Of American Agriculture 22
TRACTORS AND TRUCKS FOR WAR 24
PEACETIME CHALLENGES 30
INTO THE OIL PATCH AND BEYOND 37
THE ROARING DECADE 46
Out of Africa 47
The CFR Engine 55

DEPRESSION AND WAR 59

The Ice Engine 62
The Horning Memorial Award 75
A PIVOTAL YEAR 76
THE LITTLE DEPRESSION 80
Waukesha Goes To Sea 84
WORLD WAR II 86
War Contribution 93
Waukesha At War 94

STATIONARY POWER 97

Waukeshas At Work, 1950s-1960s 102
NEW PEOPLE, NEW MOTORS 104
High-Rise Power 110
ACQUIRING AND ACQUIRED 112
BANGOR-PUNTA YEARS 117
The Motorworks 119
Minesweepers For The Navy 123

THE DRESSER YEARS 125

OLD PLANT, NEW PLANT 126
Distributors 132
THE GREAT RECESSION 134
BIGGER IS BETTER 138
Waukeshas At Work, 1980s-1990s 142
THE NEW DRESSER 144
World Trade Center Survivors 147
IN A NEW CENTURY 148

FACING FORWARD 153

Joseph Kluth, engine builder, 1930s.

Ryan Rudnitzki, Development Engineer I and Lori Welch, Controls Engineer Sr. II, engine builders, 2006.

FOREWORD

The Waukesha Engine story is a good one. It has interesting characters, human determination and ingenuity in the face of challenge and adversity, and adventure in rugged territory. And yes, there's lots of information on our engines, too.

This is an American success story and a chronicle of a changing, growing global economy.

The pressures that have driven worldwide industrial development and change in the last century are the same ones that have shaped Waukesha Engine's history. The fact that the company not only survived, but thrived, says a lot about its ability to innovate and adapt to changing market needs. At the same time, the vision and values that drove the Waukesha Motor Company in its early years are alive and well at Waukesha Engine today.

As you read this book, look for some themes that appear again and again, from the company's early days right up to the present time: acquisitions of technology as far back as the 1920s; robustly designed products that run on fuels of varying qualities; the global nature of our business, and a commitment to innovation and adaptation in products and processes.

Each year more and more of our business comes from outside the U.S. And our most recent acquisition – technology that gives us key building blocks for power generation installations world-wide – supports that focus.

Innovation in products continues, but to survive in today's competitive, global marketplace, we are also placing greater emphasis on innovative processes such as Six Sigma and lean manufacturing. Our supplier relationships are now a core competency and we have a heightened commitment to providing employees with the skills necessary for success in a global economy.

The greater emphasis on training acknowledges that it is people who have driven Waukesha Engine's success and will continue to do so – our employees working at all levels, along with our suppliers and our channel partners.

We owe much to the previous generations whose vision, dedication and skills brought the company to this point. Now it is our turn to meet the challenge to continue to innovate and adapt and take Waukesha Engine into its second century. Future historians will judge our success.

Tom Laird
October 2006
Waukesha, Wisconsin

Thomas J. Laird, President of Waukesha Engine.

Machine Shop

Machinists in the new St. Paul Avenue shop, about 1910. Power for the machinery came from the central plant by means of overhead drive shafts, pulleys and belts. Windows supplied light and ventilation. When it began operation with fifty workers in 1909, the new plant was expected to produce one thousand motors a year.

CHAPTER ONE

OPENING DECADES

Spindletop

The Lucas gusher geysered crude oil for nine days before it was capped, giving birth to an industrial revolution based on abundant petroleum and the internal combustion engine as built in Waukesha.

IN THE FIRST MONTH OF THE FIRST YEAR OF THE TWENTIETH CENTURY, HALFWAY UP THE SLOPE OF A HILLOCK IN THE OTHERWISE UNRELIEVED FLATLAND OF SOUTHEAST TEXAS, THE MODERN AGE OF PETROLEUM WAS BORN. THE DATE WAS JANUARY 10, 1901, the place was a salt dome a few miles south of Beaumont that the locals called Spindletop. After a decade of fruitless exploration a persistent oil driller named Anthony Lucas finally hit his sweet spot. The well rumbled with subterranean thunder as natural gas, sulphur sand, drilling mud, rocks and oil stormed up the casing and blew five thousand pounds of drill pipe two hundred feet into the air. When it reached the surface the stream of oil flooded over the deck of the derrick and kept on going, gushing into a one-hundred fifty foot fountain spewing black rain onto the ground below.

The famed Lucas Gusher geysered crude for nine days—one million barrels—before someone figured out a way to cap it. The well had already produced more oil in less time than any other well in North America. And it was just the beginning. In only a few years, the hundreds of drilling rigs that sprouted like a timber-frame forest on the once barren ground of the Spindletop oil field produced more petroleum than the entire American oil industry had pumped since its beginnings in Pennsylvania in 1859.

The repercussions of the Spindletop find reverberated around the world. In a hundred different workshops—from the Daimler *Motoren-Gesselschaft* in Germany to the bicycle shop of Orville and Wilbur Wright in Ohio—engineers and mechanics who had been refining the workings of the internal combustion engine for decades learned they now had what looked like an inexhaustible supply of inexpensive fuel. Instead of the scarce, costly and ultimately inefficient fuels they had been using, they now had all the gasoline their engines could burn. The challenge now was to design and build motors—they did not call them engines—that could extract the maximum in power from gasoline, withstand the wear and tear of the high speeds at which they must run and succeed in a marketplace rapidly filling with highly skilled and motivated competitors.

In the decade after Spindletop, start-up motor makers sprouted like oil derricks at Beaumont. Ross Judson started the Continental Motor Manufacturing Company in Chicago in 1905; Milwaukee's Briggs and Stratton came along in 1908, Wisconsin Motor in 1909, to name only a few. In addition to companies that concentrated solely on motors, there were numerous automotive, marine, airplane, farm equipment, and other manufacturers who built their own gasoline fueled power plants as well. Ford, Hall, Curtiss, Deere—a whole new sector of the economy was coming to life.

The electrical power industry was also stirring, and with a beginning as dramatic as the gusher at Spindletop. The construction of a power station at Niagara Falls, New York, in the 1880s had triggered what one journalist called a national outbreak of "hydromania." In the 1900s, nearly every village, city and town that had a dam built a generating plant and distribution grid. New dams flanked by generating stations spanned rivers at remote water power sites and long-distance transmission lines carried the electricity to population centers many miles away. Where water power was absent or insufficient, coal or oil-fired central power plants were constructed; in rural areas or where back up power was required, small capacity generators were installed and powered by motors burning gasoline or natural gas.

A second industrial and transportation revolution was underway. The steam technology of the eighteenth and nineteenth centuries would be replaced by the oil and electricity-based technology of the twentieth century. The company that started life as the Waukesha Motor Company would play an important role in both of these revolutions.

Like American industry, the village of Waukesha was in transition at the turn of the twentieth century. Founded in the 1830s, and soon known as a European-style resort where visitors lingered to partake of its health-bestowing spring water, Waukesha entered the industrial age. In 1886, the Wisconsin Central railroad built a car and locomotive repair shop that brought three hundred blue collar jobs to town. The North Western and the Milwaukee Road railroads were already present and together the three lines gave the little city of six-thousand people extraordinarily good rail connections. In 1891, a group of investors led by Waukesha hardware man Conrad Haertel incorporated an iron foundry they named the Waukesha Malleable Iron Company. By the turn of the century the company had over two hundred employees turning out thousands of castings a year. Two more foundries followed Waukesha

Hydromania

The installation of electrical generators at Niagara Falls, New York, launched the electrical power industry in the United States and another industrial revolution in which Waukesha engines would play an important role.

Malleable and by 1913, the casting of aluminum, brass and iron created more wealth and employed more people than Waukesha water ever did.

The automobile came to Waukesha in 1897, when Emil Estberg, who owned a jewelry shop downtown, drove his new Milwaukee Auto home from the manufacturing plant. Estberg's four-wheeled two-seater was powered by a steam boiler that burned gasoline. He drove it regularly in summer for a few years, until replacing it with a Winton touring car. One of the most popular autos of the 1900s, the Winton had garnered national attention when two hardy adventurers actually drove one from San Francisco to New York. The trip took sixty-four days and, with attendant publicity, firmly planted the automobile in the American consciousness as not just a rich man's toy but as a serious means of transportation. In 1900, the United States had about eight thousand automobiles registered on the tax rolls. By 1912, the number had grown to nine-hundred-thousand.

Harry Levan Horning was aware of the technological revolution going on in the United States and was determined to be part of it. Born in Wauwatosa, Wisconsin in 1880, and raised in Waukesha, he was the son of Henry and Martha Horning, who had a hardware store downtown. A gifted student with parents who recognized his abilities while he was still in the public grade school, Harry was enrolled in the "modern classical" or college prep course at Carroll Academy, the high school branch of Waukesha's Carroll College. Upon graduation he entered the "scientific" undergraduate program at Carroll which, combined with his Academy studies, gave him a solid grounding in the liberal arts and sciences. Scholar, engineer, mechanic, entrepreneur, fond of art, music and a story well told, Horning personified the liberal arts ideal of a truly educated man.

Harry Horning

Harry Horning was twenty-five years old when he and his partners founded the Motor Company. Horning defined his philosophy of engine-building as "science applied commercially in the development of a technical product."

In 1905, the Winton auto was reputed to have the finest motor in America. Horning and co-founder Fred Ahrens conducted after-hours research in engine design by tearing down and rebuilding motors from Wintons and other autos dropped off for an overnight wash and polish at the Blue Front Garage. Many car owners wondered why their cars ran so much better after a "wash" at the Blue Front.

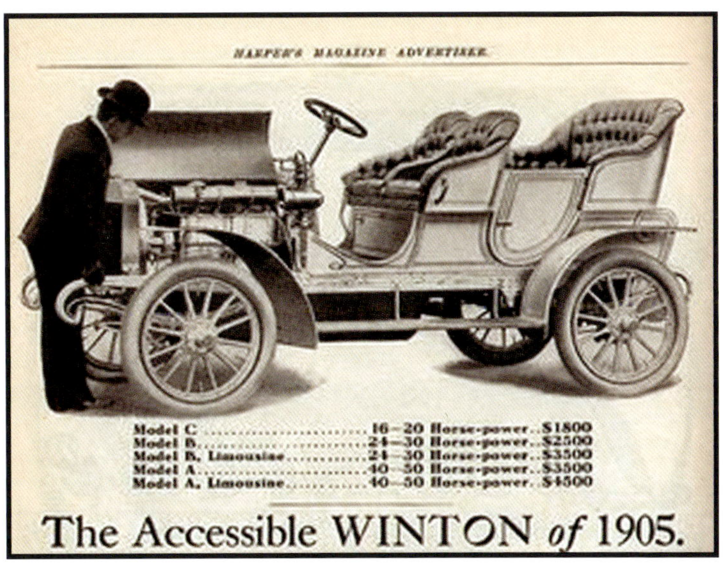

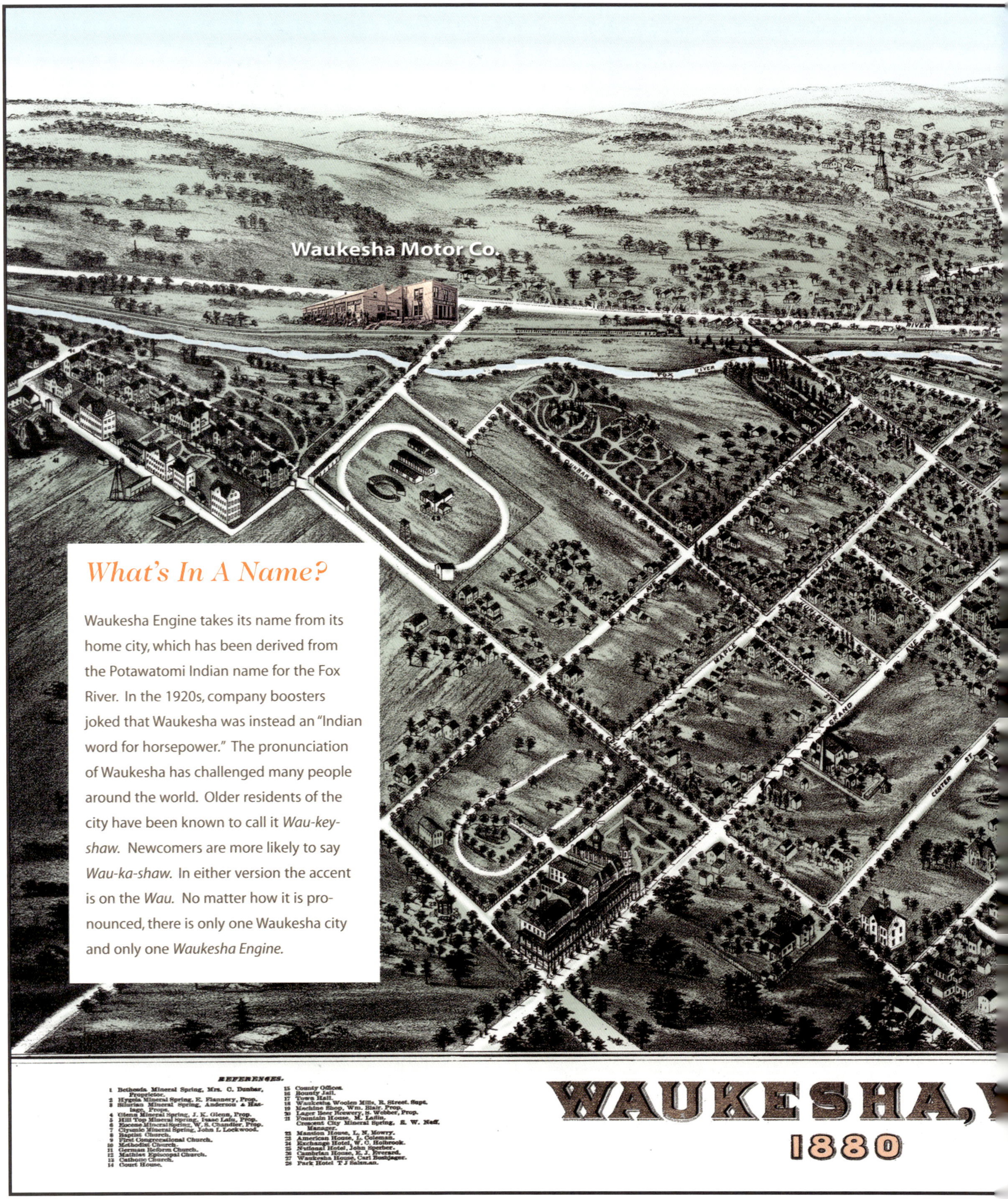

What's In A Name?

Waukesha Engine takes its name from its home city, which has been derived from the Potawatomi Indian name for the Fox River. In the 1920s, company boosters joked that Waukesha was instead an "Indian word for horsepower." The pronunciation of Waukesha has challenged many people around the world. Older residents of the city have been known to call it *Wau-key-shaw*. Newcomers are more likely to say *Wau-ka-shaw*. In either version the accent is on the *Wau*. No matter how it is pronounced, there is only one Waukesha city and only one *Waukesha Engine*.

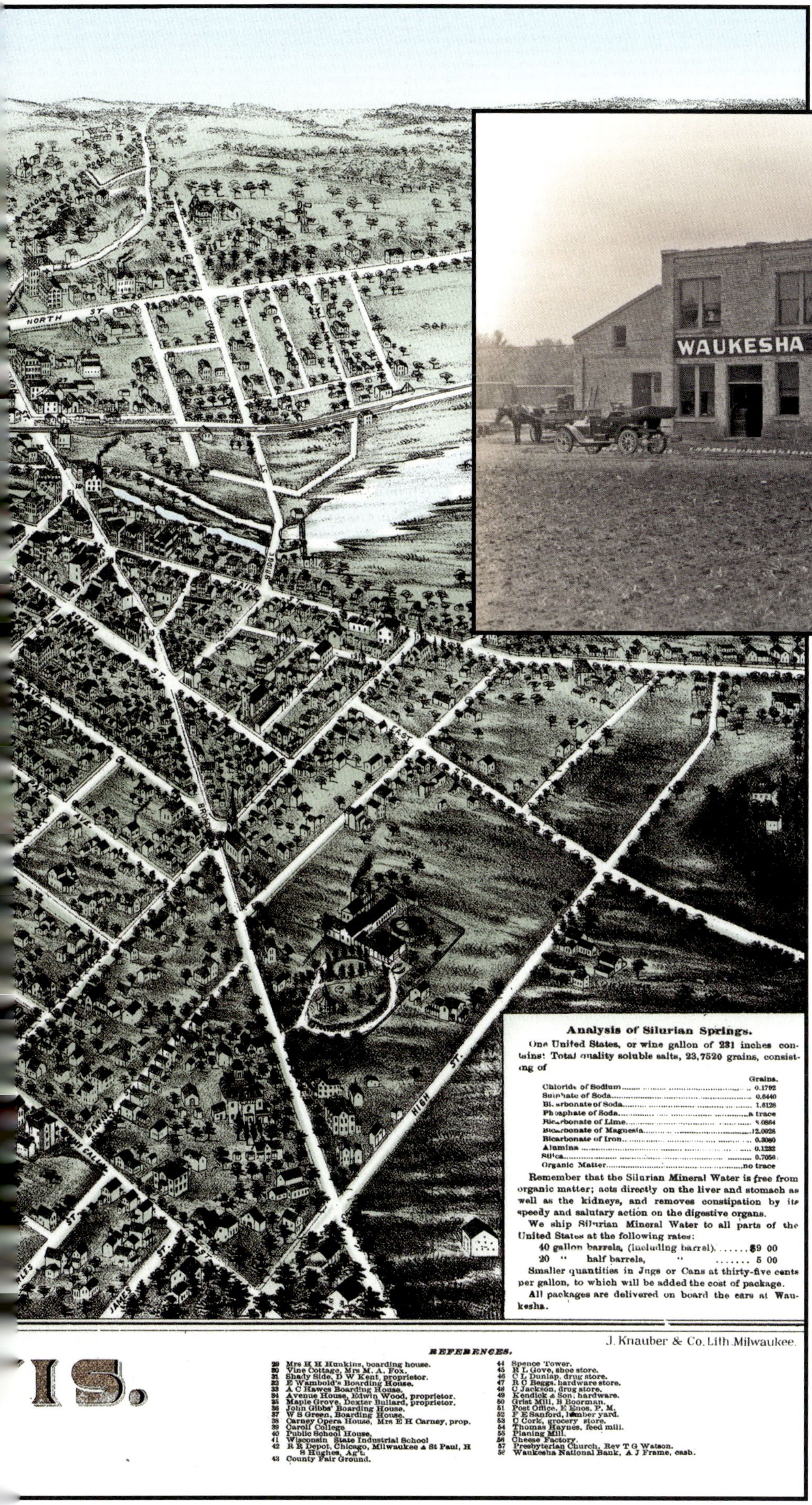

Young Village

The Waukesha Motor Company plant off West St. Paul Avenue, 1909 (above).

When this bird's-eye view of Waukesha was drawn in 1880, the city was a popular European-style resort. Luxurious hotels and humble boarding houses welcomed visitors who stayed for weeks to drink the mineral-rich water bubbling out of the city's numerous springs. The springs era was over and Waukesha was well on its way to becoming an industrial center when the Motor Company built its new plant on St. Paul Avenue. (Courtesy, Waukesha County Historical Society & Museum)

Had he been born in the 1700s Harry Horning would have felt at home in the workshops of steam power pioneers Thomas Newcomen and James Watt, or at dinner discussing the latest scientific developments with Benjamin Franklin. Had he been born a post-World War II baby boomer in the United States, Horning would have shared emails with Steve Jobs, Bill Gates and the other electronic entrepreneurs who engineered the digital revolution of the late twentieth century. As fate had it, Horning was born at the right time in the right place to make the internal combustion engine the focus of his prodigious energy and abilities.

Blue Front Garage

The Blue Front Garage on North Street was the home of Waukesha Motor from 1906 until the end of 1909.

His first job out of college in 1901 was in the chemical laboratory of the Milwaukee Gas Light Company, a behind-the-times operation that converted coal to gas for street and interior lighting. He soon moved on to the equally-antique steam engineering department at the Crane Company, also in Milwaukee. After about two years there, he came home to head up the mechanical engineering department of the Modern Steel Structural Company where he worked on the design of the Duluth-Superior aerial lift bridge, an engineering marvel in its day. At Modern Steel he struck up a friendship with Frederick Ahrens, a foreman and master mechanic who shared Horning's interest in the internal combustion engine.

The president of Modern Steel, S.B. Harding, kept a boat on Pewaukee Lake with a temperamental gasoline motor. He told Ahrens and Horning to get the balky motor running smoothly and keep it running smoothly. They took the clunker apart, put it back together, got it to run better, but recognized that no matter what they did, it would still be a clunker. If they wanted a better motor, they would have to build it themselves and that is what they resolved to do.

They opened "The Blue Front Garage" in the former office building of the Waukesha Woolen Mill Company on North Street. They would wash, store

WAUKESHA CHRONOLOGY

MAY 29, 1906
Waukesha Motor Co. incorporates, with a capitalization of $25,000 and leading investor Allan T. Stebbins as president; Fred Ahrens is vice president and Harry Horning secretary-treasurer. Ahrens and Horning build the first Waukesha Motor, the Model A.

1907
Production of a second motor, the Model H, begins and Waukesha makes its first sale to Sternberg Truck.

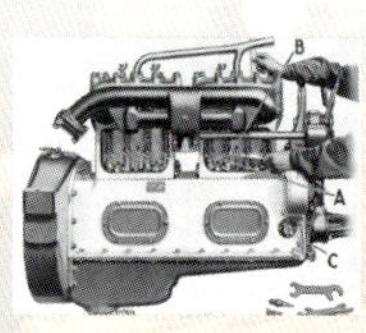

1908
Stebbins leaves the company and Samper Perkins is elected president; two more new motors, the Model I and J, are introduced.

and repair all models of automobile as well as repair and sell tires. Storing autos was an important service at the time, since car owners in Wisconsin rarely drove their cars in the winter. The weather was too cold for open air motoring and no one cleared snow off the roads anyway, so autos were parked indoors with their frames resting on blocks to take the weight off the tires. Since they kept their day jobs at Modern Steel, Horning and Ahrens needed help to mind the shop and to actually wash the cars so they hired sixteen-year-old James Remington. He would become the first wage-earning employee of the Waukesha Motor Company.

Horning had some training and experience as a mechanical engineer while Ahrens was a "mechanic," which meant he was good with machinery, but neither man had seen the insides of an internal combustion engine until they tore into S.B. Harding's boat motor. Between the cars in storage at the Blue Front and the cars they kept overnight for Remington's wash jobs, the two would-be engine builders had abundant opportunity to conduct hands-on research into the current state of gasoline engine design. After a day's work at Metal Structures, Horning and Ahrens dismantled the motors left in the garage by unsuspecting motorists. They probably tore down the engine in Emil Estberg's Winton, and if they did, they learned a lot, since the Winton was reputed to have the finest motor made in America. By the time they left for work in the morning, Horning and Ahrens had the motor back together, the oil smears wiped off the fenders and the car ready for pickup. If anyone was suspicious, no one complained. Instead, more than one car owner declared that merely leaving his car for a wash at the Blue Front improved its performance remarkably—as if Jim Remington's wash rag was really a magic wand.

After months of after-hours research, the midnight mechanics were ready to take the entrepreneurial plunge. They put together whatever capital they had between them and found one unlikely investor willing to chip in. Allen T. Stebbins was a sixty-year-old retired farmer who had recently moved to Waukesha and who, despite his years, was open to new ideas. He joined the two younger men to incorporate the Waukesha Motor Company on May 29, 1906. The initial capitalization was $25,000 (about $525,000 in 2006 dollars). Stebbins became the first president of the company, Ahrens was vice president, and Horning secretary-treasurer. As the three principal stockholders they also constituted the board of directors. Other shareholders were Samper Perkins, Mary Wilkins Mucklestone and her sister Frances G. Wilkins. The

Allen Stebbins

A retired farmer receptive to new ideas, Stebbins supplied the initial investment to organize the Motor Company. He served as president from the day of incorporation, May 29, 1906, until February 1908.

1909
The company is reorganized and recapitalized with Conrad Haertal elected president and Horning hired as general manager. A new plant is built and occupied by the end of the year.

1910
In its first full year in its new plant, Waukesha manufactures motors for auto, tractor, truck, boat, construction and industrial applications.

1911
The Company records its first profits and pays stockholders their first cash dividend.

"The Waukesha"

The four-cylinder, 35-horsepower Model A was the company's first motor. It cost $12,000 to build and sold for $500, but still laid the foundation for the company's future success.

women's shares were voted by Mary's husband, Milo Mucklestone, attorney and future county prosecutor.

Horning and Ahrens immediately began work on an engine of their design based on the lessons they had learned during those long nights at the Blue Front. This prototype, aptly named the Model A, was a four-cylinder, 35-horsepower design with the combustion chamber cast into the cylinder block. It cost $12,000 to build and therefore consumed a considerable portion of the company's capital. Since all the parts had to be specially made from Horning's and Ahrens' original designs, it was expensive, as prototypes tend to be—but it worked. The Motor Company was on its way. The first Model A sold for $500 to a boating enthusiast on Pewaukee Lake who also dabbled in sales and who quickly resold it for more than he paid.

Horning and Ahrens received more orders from boat builders in Madison and Lake Geneva. Car and truck builders in Milwaukee, Chicago, Pittsburgh and New York expressed interest. In 1907, Horning and Ahrens standardized the design and began production of two motors, the Model A and the forty-five-horsepower Model H. As a press release, probably written by Horning and published that year states, Waukesha motors were "of the well known Renault type," and possessed all the best features of the design, which included "a self-contained system of lubrication." The release went on to list other features that made the Waukesha a superior motor: "cylinders are cast in pairs…ground to exact size…pistons of the same grade of iron as the cylinders…the valves [are] amply water-jacketed…the brass water pump…oil constantly pumped from an oil-well below the crankcase…high tension ignition…" The most important feature that Waukesha borrowed from Renault and which it touted in its advertising was "the self-contained oiling system." Exaggerating only slightly, what was quite likely the company's first publicity piece concluded that, "the design of these motors is the result of long experience in this field and nothing has been omitted to make them represent as close-cut and compact unit power plants as are obtainable on the market."

Its motors were up to date and its prospects bright, but the little company was barely two years old when it faced its first management crisis. At a special stockholders meeting in February 1908, Stebbins, Ahrens and Horning submitted their resignations as officers and board members. The stockholders then unanimously elected a new board consisting of Horning, Ahrens and Sam Perkins. As directors, they then elected themselves as officers, with Perkins as president, Ahrens vice president and Horning secretary-treasurer. It is not clear why Stebbins left the company, only that he and his family relocated to Minneapolis. His involvement in the company was necessarily brief, but crucial. Without his support his young partners might not have gotten their business off the ground.

Sam Perkins was forty-three years old, a salesman and partner in a hardware store in downtown Waukesha. As such, he was a competitor of the Hornings, but their relationship was far from hostile. Perkins had worked in Henry Horning's hardware store when he first came to Waukesha as a teenager and soon became part of the family. He was unmarried and, following the custom of the time, rented bed and board in a family home, which in his case was the Hornings'. He lodged with the family for over thirty years, relocating when they did, and developed a close relationship with Harry, who was seventeen years his junior. Perkins was one of the original investors in Waukesha Motor and remained on the board until his death in 1932, but he made his greatest contribution in the year he served as president.

Perkins recognized that if Waukesha Motor was to grow it needed capital. In the course of 1908 and early 1909, he set out to find it. According to William Dunlap, who ran a drug store in downtown Waukesha, Perkins was in

Sam Perkins

Hardware salesman and close friend of the Horning family, Samper Perkins found the investors that enabled the garage operation of Horning, Ahrens and Stebbins to reorganize into a viable manufacturing company. He served as president for less than two years but held a director's seat until 1932.

Named for its passengers' favorite pastime and not its motor, the *Chatterbox* was a gasoline launch powered by an early Waukesha motor.

the store and happened to mention that he had found a pool of investors in Flint, Michigan, and was about to move the company there. Alarmed, Dunlap hurriedly summoned a meeting of local business leaders who, minutes before Perkins was about to meet with a delegation from Flint, came up with the money to keep the motor company in Waukesha. At a stockholders meeting in May 1909, what might have become the Flint Motor Company was recapitalized at $100,000 and the additional shares of stock offered to new shareholders. Bankers headed the list of subscribers: Andrew J. Frame, president of Waukesha National; Edward Estberg, cashier of the bank and brother of the Winton auto owner; Charles Nelson, Sr., loan officer at First Wisconsin National Bank in Milwaukee. They were joined by Frank H. Putney, attorney, county judge and co-founder of Waukesha Malleable; his partner, foundryman Conrad Haertel; businessmen A.S. Putney and Charles H. Newbury. Perkins, Horning and the Mucklestone-Wilkins combine also purchased more stock, with Perkins owning the largest bloc, sixty shares. To give the new shareholders a voice in the company, two seats were added to the board of directors. At a subsequent meeting, Haertel, Estberg and Nelson were elected to the board, along with Perkins and Horning. They then elected Conrad Haertel as president and hired Horning as general manager. Horning was twenty-nine years old.

Seeing the handwriting on the wall, Fred Ahrens resigned before the board election. He continued to work for the company and retained his title of superintendent with a weekly salary of $25, but it seems that the master mechanic did not fit into the new regime. He soon sold his twenty-five shares

New Plant

The Motor Company moved to its plant on the outskirts of Waukesha in 1909. Since shipping was exclusively by rail, a location along the tracks was the prime consideration. It also gave the company a chance to tell rail passengers that they were passing "THE PLACE HIGH GRADE AUTOMOBILE MOTORS ARE MADE." The words were painted on the first shop addition, already necessary and completed in 1910.

of stock, worth $100 each, and left the company and the city. In 1906, the proverbial two men and a boy—Fred Ahrens, Harry Horning and James Remington—built the first Waukesha motor. Three years later, Sam Perkins brought in the capital and management experience that built the Waukesha Motor Company.

A NEW COMPANY

Among its first actions the new board authorized a search for a site to build a new plant. Two requirements had to be met. The site had to be large enough to accommodate future growth and it had to be on or very near a railroad line so goods could be conveniently shipped in and out. A search committee headed by Charles Nelson found the right property on the undeveloped west side of Waukesha along what was then West St. Paul Avenue. The avenue took its name from the Milwaukee Road rail line it paralleled, which was also known as the "St. Paul." The acreage the company purchased for $3,000 included ground on both sides of the Milwaukee Road tracks and was only a few hundred feet from the point where the Chicago and North Western tracks crossed the St. Paul. For quick and easy rail access, it was hard to beat.

Group Photo

Company personnel, and a four-legged guest, gathered for the photographer, 1910. Twenty-nine-year-old general manager Harry Horning is seated first row center holding his straw hat. Sam Perkins leans over his left shoulder. In the top row, fourth from the left stands James Remington, the first wage-earning employee of Waukesha Motor.

Model R

Popular with construction, farm equipment and truck manufacturers, the Model R was introduced in 1913.

The Model R and other engines of the pre-World War I years enabled Waukesha to pioneer the use of gasoline engines in concrete-mixing and paving equipment in 1914.

Ironically, for a gasoline motor manufacturer, highway access was less important than rail service. The new plant was built two-hundred and fifty feet south of West St. Paul Avenue. The company graded its own road, appropriately called Factory Street, to run from West St. Paul to the front door of the plant.

The Motor Company's closest neighbor in its new home was a relic of the old Waukesha. On the hill across West St. Paul sat the Keeley Institute, where "Dr. Lester Keeley" administered his "marvelous remedies…for the cure of liquor, opium, chloral and tobacco habits." In 1913 a representative of the new city arrived when the Spring City Foundry was built across Factory Street. It eventually filled most of the space east to State (Prairie) Avenue.

The board budgeted $10,000 to build the new plant and the lowest bid came in at $10,218, close enough to be accepted. The contractor was a five-year-old Waukesha company known as Advance Construction, which billed itself as a specialist in concrete construction. Like the Motor Company, Advance Construction was a purveyor of new technology. It was taking advantage of improvements in the manufacture of Portland cement that had dramatically increased the supply and lowered the cost of concrete. The age of oil began in the 1900s and so did the age of concrete. They went hand in hand. Cars and trucks needed paved roads and concrete supplied the pavement. As a builder of engines for the automotive and construction fields Waukesha Motor was in the new game right at its start.

The Waukesha Motor plant was an example of how inexpensive and abundant concrete affected building design. Instead of laying stone footings and wooden decking as it would have in the 1880s, Advance Construction poured concrete, the cutting-edge building technique of the 1900s. The new plant consisted of a two-story brick office building, twenty-four by forty-eight feet, with offices downstairs and storage on the second floor, and the factory stretching west along the railroad tracks. It consisted of four steel-framed

units with brick exteriors, each one measuring twenty-two by eighty-eight feet and covered by a saw-tooth roof with windows facing north to illuminate the shop floor below. Electricity to run the machinery came from the new power plant of the Waukesha Gas and Electric Company. No sooner was it done, than it proved to be too small. In the summer of 1910, a fifth unit was added. It was the first of many additions built over the next twenty years. Growth was constant until 1929 and rare was the summer when the concrete mixers were not enlarging the plant at Waukesha Motor.

In December 1909 fifty workers moved into the new plant, which was reportedly designed to produce at least one thousand motors a year. It was an optimistic projection, since the company shipped a total number of ninety-nine motors out of the Blue Front Garage in 1909. In addition to the Model A and H, Waukesha had introduced the Model AU4, with enclosed valve springs and timing gear; the 30-horsepower Model I; and the 50-horsepower Model J. List prices ranged from $330 ($6,800 in 2006) for the best-selling Model I, to $500 for the Model J. Customers included just about anybody who walked in the door—so to speak. Auto and truck builders, boat manufacturers, farm and construction equipment makers were many in number and they were all looking for power plants, yet few were able to place large-volume orders. A typical order was for the seven Model Hs sold to the Harder Truck Company for the city of Chicago's first motorized fire trucks in 1911. Consequently, Waukesha shipped a few motors here, a few motors there, without nailing

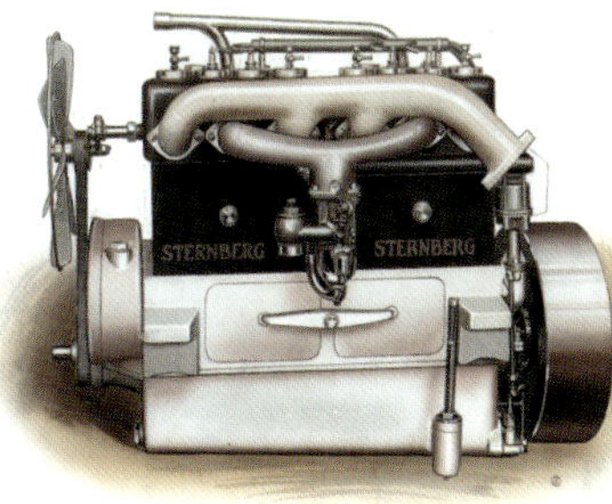

Model M

A Model M manufactured for Sternberg Truck before it changed its name to Sterling in 1916. Introduced in 1912, the Model M was one of the first two Waukeshas to cross the ten thousand sales threshold.

A Sternberg truck filled with motor makers rode in a parade in Waukesha, probably in 1908. Third from left is Harry Horning, fifth is Sam Perkins, and ninth is Fred Ahrens.

Fire Power

This truck was powered by one of seven Model H motors Waukesha sold to Chicago's Harder Truck Company in 1911. It has been recognized as Chicago's first gasoline-engine-powered fire truck. Waukesha made its first sale of an engine for fire patrol duties to the city of Chicago in 1908.

down the bread-and-butter contract that would send a steady flow of a single model out the door and bring regular cash back in return. Dictated by necessity at first, adaptability and innovation to meet the demands of diverse and dynamic markets became standard practice at Waukesha.

Harry Horning maintained the commitment to research and development that had kept him and Fred Ahrens awake all night at the Blue Front Garage. Seven new models were introduced between 1909 and 1911, the most popular of which was the AU4. In addition, the groundwork was laid for the big sellers that first appeared in 1912. Horning, who did not have an engineering degree, also took pains to establish his professional credentials. Already a member of the American Society of Mechanical Engineers, he was elected to the Society of Automotive Engineers in 1910. He was recognized for his work in developing motors for farm tractors and for his continuing interest in improving the quality of the fuel burned in gasoline engines. He also took time out to get married. Harry Horning and Elsie Muir were wed in June, 1910. They made a two-week honeymoon tour of the eastern states before Elsie settled into homemaking and Harry went back to making motors.

Waukesha MOTORS

1912
Company publishes its first in-house newsletter, *Power and Efficiency.*

1913
The Model R is introduced, one of Waukesha's first 10,000+ sellers.

1914
World War I begins in Europe, stimulating demand for farm tractors and trucks powered by Waukesha motors.

The company balance sheet was registering its first profits. A news reporter visiting the plant in December 1910 recorded that one contract with a "leading motor boat concern" was bringing in $16,000 a month and the material in the factory, prior to assembly into motors, was worth $141,000. At the end of the fiscal year in June 1911, Horning reported that, "a net profit of $20,000.00 has been earned by the company." Previously, any income above expenses had "been invested in permanent additions to the company's property" but now it was finally time to reward the investors with a dividend "to the extent of $20,000.00" in stock. Fiscal 1912 was also profitable and the shareholders received a stock dividend of $13,560. General Manager Horning's salary was raised to $250 a month, plus a bonus of eight per cent on all profit above $25,000.

Fiscal 1912-'13 was the break out year for Waukesha Motor. The 42-horsepower Model L and Model M; the 51-horsepower Model P; and the 35-horsepower Model R, made their debut. They were the engines that carried the company through World War I and into the 1920s. They all broke the one thousand unit sales mark, with the M and the R becoming the first motors to cross the ten thousand sales threshold. The Company also connected with two of its most important and faithful customers. Sternberg Motor Truck, which had purchased its first Waukesha motor in 1907, made what became a four-decade-long commitment to the company in 1912. With its name anglicized to Sterling, the Milwaukee truck builder continued to buy Waukeshas until it was acquired by White Motor Company in 1951. In 1913, Waukesha received its first purchase order from Ingersoll-Rand, also for a Model L. Thanks to Ingersoll, Waukesha became a pioneer in the manufacture of gasoline motors used to run air compressors. Ingersoll became the Motor Company's largest construction equipment account and would continue to purchase Waukeshas every year until 1955.

Air Power

A Model L sold to Ingersoll-Rand in 1913 inaugurated a relationship that lasted over forty years. Together Waukesha and Ingersoll-Rand pioneered the manufacture of portable compressed-air units. The unit depicted here dates from the 1930s.

1915
Expansion underway that will double the size of the plant in a year.

1916
The Model T is introduced; it is the first Waukesha with detachable cylinder heads.

1917
US enters World War I and the company participates in the national program to produce military trucks. The first truck delivered to President Wilson is powered by a Military Class B motor built at Waukesha.

The Golden Age of American Agriculture

1

The farm economy prospered in the first twenty years of the twentieth century, encouraging American farmers to invest in new equipment and prompting a flood of new tractor manufacturers to enter the market. Waukesha built motors for over one hundred tractor makers from the 1910s until the 1970s. Most did not survive beyond the 1920s. The Models L, M, N, P and R were the most popular tractor engines in the years prior to World War I and the largest customers were the Rock Island Plow Company and Huber Manufacturing.

1. The one-of-a-kind Nevada Auto Plow, built in Nevada, Iowa.

2. A "Steel-Mule" tractor from the Joliet Oil Tractor Company, Joliet, Illinois.

3. A Russell "Little Boss" tractor. A decades-old steam engine builder from Massillon, Ohio, Russell switched to gasoline power in the 1910s. Russell touted itself as the "Old Reliable," as symbolized by the distinctive Holstein bull logo painted on its tractors.

4. A 1920s Huber three-wheeled "Modern Farmer." Located in Marion, Ohio, Huber was among the first tractor builders to order hundreds of Waukesha engines.

2

3

4

5

5. Lawter Tractor built three-wheeled, three-plow, Waukesha-powered tractors in St. Mary's, Ohio.

6

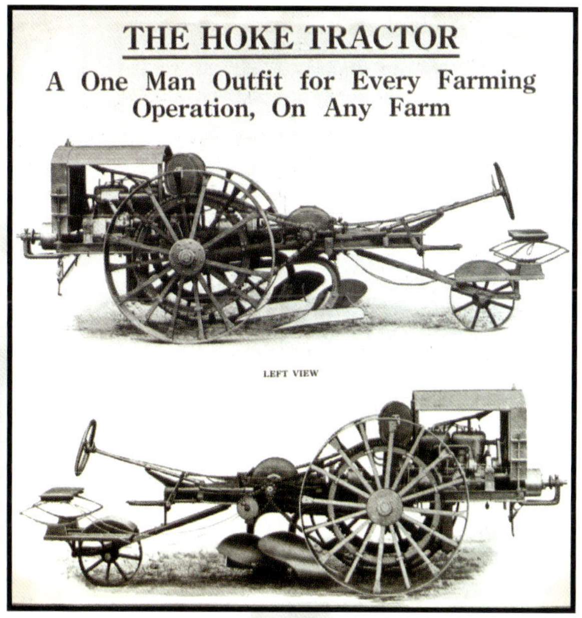

7

8

9

Modern Farmer

6. The steering system of the Detroit "Mechanical Horse" was designed to be user-friendly for farmers accustomed to horses and reins.

7. Hoke Tractor, of South Bend, Indiana, used a modern steering wheel but relied on a wagon seat suspension system for operator comfort.

8. A four-wheeled Huber "Modern Farmer" from the 1930s.

9. A Samson "Sieve-Grip" three-wheeler. Founded in Stockton, California, Samson was acquired by General Motors in the early 1920s and moved to Janesville, Wisconsin.

Liberty Truck

The Model B Liberty truck was the first American-built transport vehicle commissioned for duty in World War I. With the Model B's engine design committee headed by Harry Horning, it represented a personal triumph for Waukesha's president and for his company.

A carefully worded pre-World War I ad enabled the young motor company to claim a top spot in engine manufacturing.

TRACTORS AND TRUCKS FOR WAR

The first two decades of the twentieth century have been called "the golden age" of American agriculture, a prolonged period in which the balance between price and production consistently favored the farmer. Incomes increased, living standards improved and farmers were willing and able to invest in the new machinery offered by manufacturers who seemed to have sprung up in every city, town and village in the agricultural heartland. The city of Waukesha itself had at least two, albeit short-lived, tractor manufacturers in these years. Since the majority of Americans still lived on farms or in small communities tied to agriculture, good times on the farm meant good times for the nation.

The good times got better when the outbreak of war in Europe in August 1914 spiked demand for American agricultural products. With the models it introduced in 1912-13 and the improved versions that came along in the years immediately following, Waukesha was well-positioned to compete in the burgeoning farm tractor market. It was a welcome addition to the comfortable niche the company was creating for itself in the construction equipment and automotive truck markets. In 1914, Waukesha shipped six-hundred-twenty-eight motors, a five hundred percent increase over 1909. More than five hundred of the motors shipped in 1914 were Models L, M, P and R to be installed in tractors and trucks.

By October 1914, it was apparent that the European war was not going to end soon, so the board of directors recognized that "the prospects for a very profitable foreign and domestic business are assured to us in so large a volume, in the near future, that our present facilities in building, machinery and equipment are inadequate to cope with deliveries with sufficient promptness." The board then authorized the first of a series of additions during the World War I period that would double the size of the plant in 1915 and double it again in 1916.

A single shop addition and $1,800 in machinery were ordered in late 1914. Twenty-one additional shop sections were ordered in early 1915, plus two more sections for a "new and permanent Testing room", plus a power plant to heat the factory and run the $5,000 worth of new machinery on order. By the end of the year, bids were requested for an additional fifty-eight sections for motor assembly, testing, storage of rough stock and for shipping and receiving. Because of the "unusual conditions in the machine tool market in which it is increasingly difficult to get machinery," the board authorized Horning to spend $15,000 to "purchase for immediate and future delivery machinery whether second hand or new." In 1916, the testing room was doubled in size by the addition of four bays. A twenty-four-section machine shop came next, along with a six-section carpenters' and pattern shop. In the midst of the expansion in September 1915 the *Waukesha Freeman* reported that the plant was turning out motors at the rate of five thousand units a year and would ship fifty percent more than five thousand in 1916. Farm tractor manufacturers headed the list of customers, with Huber Manufacturing in Marion, Ohio and the Rock Island Plow Company in Illinois, manufacturer of the Heider Tractor, ordering Model M, N and R motors by the thousands.

Over Hill, Over Dale

On its way to Washington, D. C., to be presented to the War Department, one of the first two Model B trucks stopped to help another vehicle out of the mud. For reasons of security in wartime, the location of the plant that built the truck was identified on the banner it bore as "somewhere in N.Y."

Yes, Mr. President

Harry Horning listened attentively to President Woodrow Wilson when he presented the first Model B to the government.

To finance growth at this dizzying pace, the directors authorized the sale of an additional ten thousand shares of stock at $100 per that would raise one million dollars ($10.8 million in 2006). The company that had started with $25,000 and one motor had, in ten years become a multi-million-dollar operation, with several acres of manufacturing plant under roof, thirty-seven engine models in production, and one thousand employees on the payroll. The rising tide at the Motor Company lifted other local businesses with it, especially the foundries that supplied castings and forgings. Despite the halt in European immigration precipitated by the war, so many new workers had moved to Waukesha that the population had increased from 8,740 in 1910 and was on its way to 12,558 in 1920. Housing was in short supply and to meet this need, Sam Perkins and other local businessmen organized the Manufacturer's Building Company to construct and finance reasonably priced housing for blue-collar workers.

In April 1917, the United States entered the war in Europe on the side of Britain, France, Russia and their allies. Their leaders believed that the United States could best aid the war effort by supplying men to replace the millions of soldiers they had already fed into the slaughterhouse of trench warfare. The British and French would equip the Americans and integrate them into their armies. American military commanders, led by General John Pershing, disagreed. They insisted on an independent force that would cooperate with the allied armies on an equal basis. At the same time, American business leaders insisted that an American army should be armed and equipped by American manufacturers. Their patriotism and nose for business blinded them to the fact that nothing approximating a defense industry existed in the United States. Particularly when it came to equipment for modern warfare as it had evolved since 1914—heavy artillery, submarines, aircraft and military transport—the United States ranked near the bottom of the industrialized world.

Like the army and navy, American industry had to mobilize for the war and competitors had to agree to cooperate in the national interest. Instead of competing to sell their own trucks to the army, for example, as they would have preferred to do, automotive manufacturers were ordered to collectively design a few new vehicles that all of them would produce and deliver as ordered. It was meant to be a fair, simple and corruption-free means to meet an urgent national crisis. To oversee this operation, which included all branches of manufacturing for the war, President Woodrow Wilson established the War Industries Board. It was able to arm-wrestle the nation's truck builders into cooperating by ordering one hundred thousand current model trucks, pledging to ship overseas any trucks delivered and to not bring them back for sale in the United States after the war.

The Board created a set of committees and sub-committees for each category of manufacturing. Harry Horning was named Chair of the Design Committee of Engineers and commissioned to draw plans for the motors to be used in what came to be known as the Liberty Class B 3.5-ton truck and the larger Liberty Class AA 5-ton truck. That he was named chair of the committee bespeaks the esteem in which his fellow engineers held the man from Waukesha. Their confidence was justified. Along with his other qualities, Horning was very good at getting people to work together. One of his favorite management tools was a simple set of cartoons depicting a pair of mules that can't budge their load because they are pulling apart, but trot right along when they pull together.

Horning was able to get the mules from the Continental, Buda, Hercules, Wisconsin and other motor companies on the design committee to pull together and draw plans for an engine in about a month. Waukesha's vice president of engineering, J. B. Fisher, was on the design team. After settling on the design, they each agreed to focus on manufacturing only a few parts of the engine and then sharing them for the final assembly—at least for the prototype model. Waukesha was responsible entirely or in part for the cylinder

First In War

The Model B engine was mounted on a parade float celebrating the end of World War I.

War Workers

A portion of the World War I shop crew lined up for the camera on the railroad track side of the building. To meet increased wartime demand, the Motor Company enlarged its work force to one thousand men and women.

heads, crankcase, oiling system, roller pushrods, governor, intake and exhaust manifolds. The casting for the crankcase came from the Werra Foundry in Waukesha. In not quite twelve days from the time the blueprints arrived at the plant, Waukesha completed the first Class B engine. With its 4.75-inch bore and 6-inch stroke, it shared some of the characteristics of earlier "long-stroke" Waukesha models. Despite the fact that it was designed by a committee of rivals, it performed very well.

The engine was ready in September 1917 and two entire trucks were completed in October. They were driven from their assembly plants in Lima, Ohio and Rochester, New York to Washington D.C. Horning accompanied the New York model to the capital and made the official presentation to President Wilson. He was rewarded for the design committee's success with a promotion to Chair of the entire Automotive Products Section of the War Industries Board. The truck was less successful. After further testing, the army concluded that the 3.5-ton Class B and the Class AA were not suitable for conditions in France and ordered a redesigned 3-ton Class B. This vehicle was a better fit and the army ordered eight thousand units, but not until May 1918. Sixteen truck manufacturers started production, but by the time the war ended in November, the army had taken delivery of no more than twenty-five hundred. Waukesha Motor received its first order for two thousand Class B motors in November 1917, along with a $350,000 ($4.4 million in 2006) down payment. It was the largest single order for motors the company had yet received. More and larger orders were expected since no one thought that the war would end in a year. The Waukesha board authorized the purchase of

$50,000 worth of new machinery and again enlarged the plant. Another test room was added, along with fourteen additional shop sections for heat treating, blacksmithing and metallurgy, six more sections for a new boiler room and coal storage, and a two hundred-foot-long shed attached to the wall of the Spring City Foundry for workers at both companies to park autos and bicycles. The day had yet to come when all the people who built parts for autos could afford to own one. Nonetheless, the volume of motorized traffic on the factory grounds had grown so heavy that the company hired contractors to pour twenty-thousand square feet of concrete to pave storage areas and roadways.

The pressure on Henry Horning must have been intense. He was attempting to serve his country as a "dollar-a-year" man who volunteered his time and expertise to build a vital defense industry and also manage a growing company that had become an important supplier of equipment to that very industry. He spent days out of the office, traveling by rail to meet with War Department officials in Washington or with other manufacturers in midwestern and northeastern states. In May 1918, the board resolved that it "was desirable" for Horning to relinquish his government post and "give all of his time to the rapidly growing business." He agreed with the reservation that if an emergency arose and his country called, he would answer. Fortunately for Horning, Waukesha Motor and the whole world, the war ended in six months. Waukesha's general manager was now as well known in the motor manufacturing industry as any man in the country. In financial terms, the war had been good to the company. Even with the tremendous investments in plant and machinery made since 1915, profits had been large enough for the shareholders to receive more than $500,000 in dividends paid in cash and stock in the fiscal years ending in 1917 and 1918.

War Test Room

The engine test room during the World War I era.

PEACETIME CHALLENGES

A world at peace presented a new set of challenges. The Motor Company had increased its capacity many times over in only a few years, but so had its competitors. How were they all to utilize their expanded plants and enlarged work forces? As the war had demonstrated with its motorized cars, trucks, tanks, aircraft and stationary equipment, opportunities abounded for new applications of the internal combustion engine. What was the best course to follow? Motor to make? Market to pursue? For Harry Horning the answer would be found where it always had been, in searching for the information necessary to produce the motors that would compete successfully in a world of new markets.

With the war over, transatlantic travel resumed and, like thousands of Americans, Harry Horning went to Europe. He and Elsie visited the popular tourist sites, but Horning was interested in more than Buckingham Palace and the Eiffel Tower. He trekked sixty miles out of London to Shoreham, a remote spot on the coast of West Sussex to the new Bridge Works plant of Harry Ricardo. A thirty-year-old engineer and scientist, Ricardo had overcome the perception that he was just a fuzzy-minded intellectual by designing a compact (for the times) but powerful six-cylinder, 150 horsepower engine that powered the British army's Mark V tank. With eight thousand units shipped, Ricardo's engine was the first mass-produced motor in Britain and, when installed in the Mark V tank, a key contributor to the Allied victory in World War I. One German general stated that his army had not been defeated by any Allied commander but by "General Tank."

At the end of the war, the Ricardo engine represented the state of the art in automotive power plants. Its most distinctive feature was its cylinder head, with its enlarged combustion chamber that increased the "turbulence" of the fuel mix and thereby boosted the speed and efficiency of the combustion process. As his biographer observed, Ricardo's cylinder head exemplified his ability "to combine science and engineering."

Tall, lean, reserved, Harry Ricardo was to all appearances the opposite of the short, pear-shaped, gregarious Harry Horning. But when they discussed each other's work, their shared scientific approach to practical engineering, their mutual interest in perfecting motors and fuels, they were as alike as the names they shared. The two Harrys became fast friends, corresponded regu-

The Ricardo Head

The Ricardo cylinder head made engines more efficient and powerful. When it obtained licenses to build the Ricardo head, Waukesha made its first acquisition of foreign technology.

1918
World War I ends and Waukesha's contract for 2,000 military truck motors is cancelled.

1919
Harry Horning travels to England to visit famed automotive engineer Harry Ricardo.

1920
Waukesha receives its largest order to date: 10,000 Model HC engines for International Harvestor.

Thinkers And Doers

The two Harrys--Ricardo and Horning—shared their passion for innovative engineering and developed a close corporate and personal friendship.

1921
The economy is in recession; industrial production falls 30%, unemployment approaches 12%.

1922
Harry Ricardo visits the plant and discusses licensing his patents to the Motor Company.

1923
James DeLong is named Oilfield Sales Manager and makes the first sales of Waukeshas to the American oil industry.

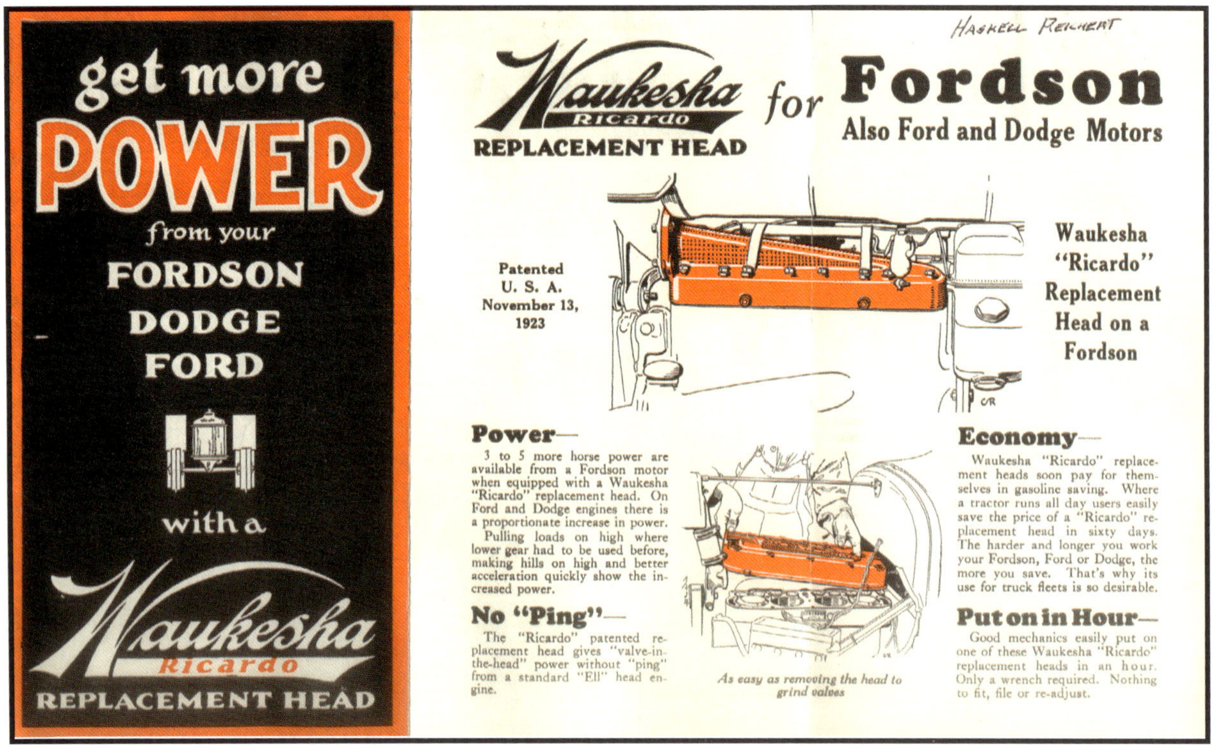

More Power To You

Displays for Ricardo replacement heads. In addition to mounting Ricardo heads on its own engines, Waukesha developed replacement heads for motors from other manufacturers.

larly and, for the next decade and one-half, visited each other as often as two busy schedules and the stately pace of ocean travel allowed.

Before he met Horning, Ricardo had sent a representative to the United States to obtain an American patent and sell licenses for his turbulent cylinder head to American manufacturers. The United States Patent Office refused to grant Ricardo a patent, so Ford, General Motors and other automotive manufacturers simply adopted his design without bothering to acquire a license.

In 1921, Ricardo gave up on trying to obtain an American patent, but granted Harry Horning permission to patent his designs in the United States in Waukesha's name. The motor company would then collect royalties from manufacturers it authorized to use the turbulent head and send Ricardo a share. Horning obtained the patent for the Ricardo head and later for a diesel engine his British friend also designed, but could not protect either one without expensive legal proceedings. As a result, Waukesha Motor acquired little more than an exclusive right to advertise the "Ricardo Head" on its motors while other manufacturers used the head without advertising it as such or acknowledging its inventor. Starting in 1923 and well into the 1930s, Waukesha Motor touted the Ricardo Head as the key component that made its engines better than its competitors'. The company also made Ricardo type replacement heads for other engine builders including American La France fire trucks, Fordson and John Deere tractors, and Oakland, Studebaker and Whippet autos. In homage to Ricardo's concept, the management newsletter that Waukesha started publishing in 1926 was entitled *Turbulence*.

Horning was already planning innovations of his own before he met Ricardo. In 1919, Waukesha introduced a new line of higher speed engines—the 42-horsepower Models BUX and CU; the 50-horsepower Model DU; the 59-horsepower Model EU; and the 37-horsepower Model FU—and discontinued production of many older models. The U suffix meant the engine was sold as a unit that included a radiator, power takeoff or—for stationery applications—protective housing. Advertised for "Trucks, Taxicabs, Speed Trucks, Tractors, Industrial Machinery," the new motors were in full production in July 1920.

The shop was humming with new production that month, but a discordant note was heard in the boardroom. Horning announced that he had been asked to take a seat on the board of a small motor parts company. Recalling what it had been like when Horning was absent during the war, the board members present—Haertel, Estberg, Perkins—expressed negative sentiments. Horning then offered his resignation to be accepted at the pleasure of the board. It must have been quite a shock to these traditionally-minded businessmen who believed firmly that the general manager's job was to mind the shop and not distract himself with outside interests. Not only were they a generation older than Horning, they had not shared his experience in the national and international motor industry. In this sense, Horning was a precursor of the corporate chief of the late twentieth century who was expected to accept additional business and community responsibilities. His board may not have understood him, but they wanted to keep him. At their next meeting the directors passed a resolution to immediately increase Horning's annual salary to $12,500 ($188,000 in 2006) and make it retroactive for one year. He no longer talked about resigning.

A Better Idea

A Fordson tractor with a Ricardo replacement head at work in Waukesha. Henry Ford had turned down Harry Ricardo's offer to sell an exclusive American license for his cylinder head, so Fordson tractor owners who wanted to improve their Ford motor installed a Waukesha replacement.

Model BUX

The BUX was the first in the line of "U" engines that led sales at Waukesha in the immediate post-World War I years.

Instead, the general manager and everyone else at the motor company talked about the International Harvestor order. The giant truck, industrial and farm equipment manufacturer asked Waukesha to supply ten thousand Model C motors for its new line of farm tractors. The contract called for Waukesha to deliver the first one hundred of what would be called the Model HC International motor in March, 1921. Production would ramp up to fifty a day in August and sustain that pace until March, 1922. Waukesha's estimators calculated that the company could make ten thousand Model HCs for $3.92 million. Adding its standard ten percent markup, Waukesha would bill International $4.35 million, plus another $500,000 for repair parts in the future. In addition, International said it was looking at producing twenty-eight thousand tractors in two years and they would all need motors. It was a very tempting deal, especially since the outlook for 1921 did not look very bright. When the International contract was discussed in September 1920, Waukesha had work for the rest of the year, but no signed contracts for 1921. Yet the company did not automatically jump into International's lap.

In the boardroom discussion, Horning pointed out that taking "this amount of business" from International "will disturb the peace of mind of our other customers." How could Waukesha serve them, they would wonder, if it was so closely tied to International? He also said that producing such a large number of motors for a single company, "would be a marked departure in our

Trucks And Buses

A Stoughton truck with a bus body and a 1920 Model BUX under the hood is on display at the Evergreen Aviation Museum, McMinnville, Oregon. Stoughton was one of over one-hundred-twenty-five truck manufacturers who used Waukesha engines. (Courtesy Evergreen Aviation Museum)

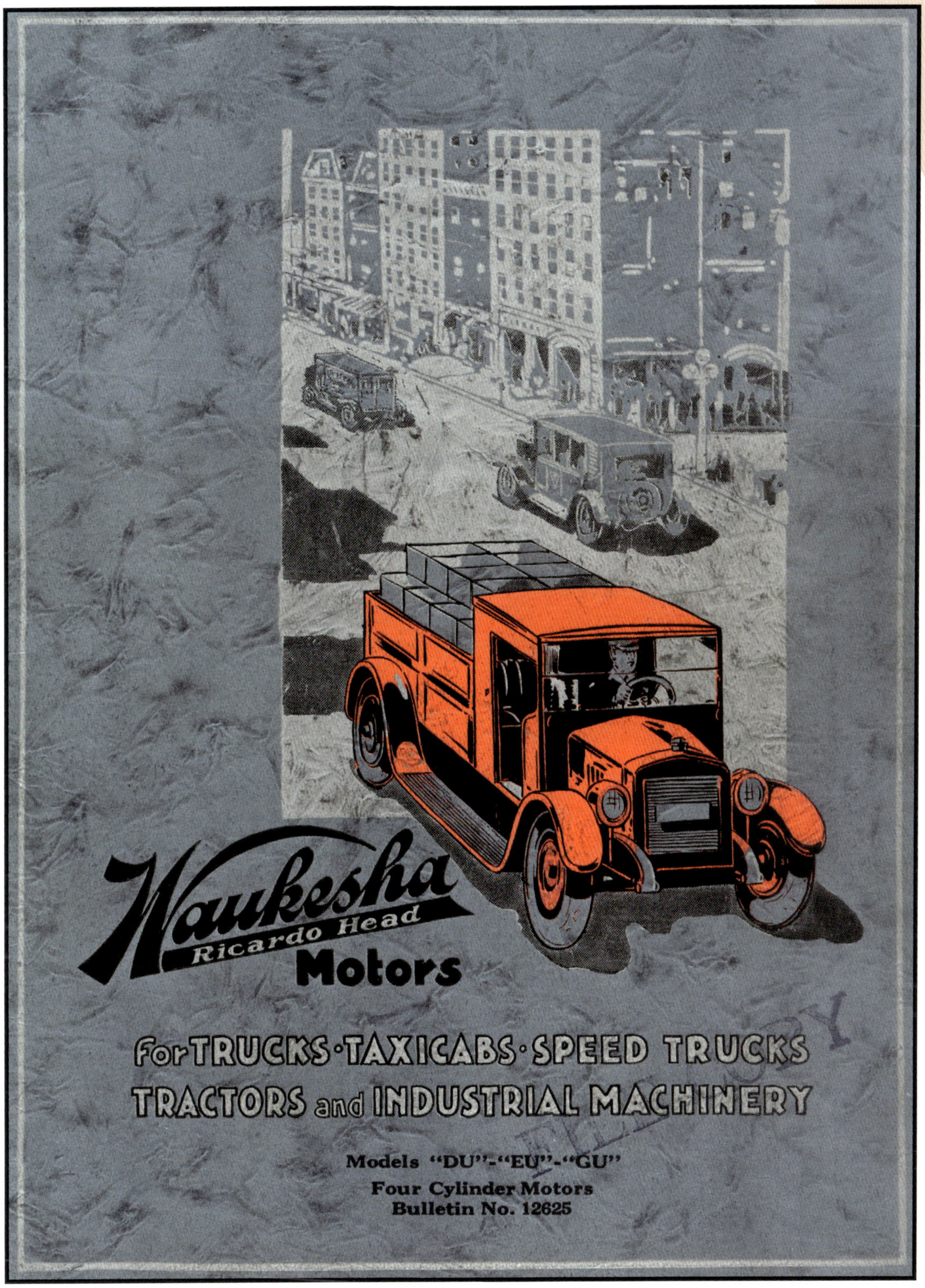

New For 1924

Sales brochure for the 1919 "U" engines, modified with Ricardo Heads for 1924.

policy, namely, eliminating the identity of the product with the purchasing public." The Waukesha imprint on the engine was important for the long-term success of the company. It made a statement saying that no matter what the application, the Waukesha engine was a separate, identifiable and superior component—not an anonymous cog in the wheel that any motor manufacturer would produce. Nonetheless, the directors approved the deal.

The uncertain state of the economy was a factor. The United States had not experienced a serious economic downturn since 1893. Harry Horning was one of many corporate managers who had never in his working life encountered a recession. That gap in his experience was about to be filled. The first sign of the times appeared with the cancellation of government contracts early in 1919. In 1920, a struggling Watson Products Company offered to pay its bill for motors and materials with stock certificates, since it did not have cash. Yet another indicator came from Huber Tractor, which had purchased nearly ten thousand motors since 1916, but placed no orders for 1921. Then Ingersoll-Rand, whose purchases had been growing steadily and had passed the eight hundred mark in 1920, ordered only one motor for 1921.

Conrad Haertel

Foundryman Conrad Haertal served as president of the Motor Company from 1909 to 1925.

Waukesha proceeded with the International order, investing $70,000 in new machinery and $20,000 in preparation work. The economy continued to slide. Republic Motor Truck stopped taking calls and answering letters, leaving a large quantity of materials and unpaid invoices with Waukesha. Inevitably, the boom fell on International. The tractor maker was forced to cut its Waukesha order in half and stretch out the delivery time for motors to the middle of 1923—when International was supposed to be on its way to building twenty-eight thousand tractors. An order for five thousand tractor motors was better than an order for none, so Waukesha and International renegotiated their contract. In 1924, after the recession ended, International ordered the first of the one hundred thousand Model XA engines it would purchase in the coming decades.

Although International and Waukesha emerged from the recession of 1921 scarred but intact, the American farmer did not. The golden age was over and the farm sector began a slow but steady slide before hitting bottom in the early 1930s. Following classic economic theory, the downturn accelerated the culling of the weakest and consolidation into the strongest due in the farm tractor industry. By the end of the decade nearly two thirds of the tractor builders operating in 1920 were out of business. The postwar recession and the persistent troubles in the farm sector only reinforced Waukesha's long-standing commitment to design new motors and find new applications for them.

INTO THE OIL PATCH AND BEYOND

James DeLong was an Indiana farm boy, born in 1889 and educated at a one-room country school. An excellent student, he was admitted to Purdue University where he obtained a degree in engineering in 1912. He started his professional career at a motor company in Indiana, and volunteered for duty when the United States entered World War I in 1917. He served in the ammunition transport section of the Army's First Infantry Division and was honorably discharged with the rank of captain. Returning to his home state, he became plant manager of the Indiana Truck Company. Indiana Truck was struggling in the recession and when Harry Horning offered him a job at Waukesha, DeLong was pleased to accept.

Horning presented his new hire with a challenging assignment: carve an entirely new market niche for Waukesha motors. The company had focused on motive power plants—tractors, trucks, buses—and construction equipment—concrete mixers, cranes, compressors. The time was right to find a new application in the industrial sector and Horning knew where he wanted DeLong to look.

Truck and tractor motor sales agents in Texas had noticed that oil drillers were still using steam engines on their rigs. They had always used steam and,

Oil Power

Workers loaded a small Oilfield Power Unit for shipment south, late 1920s.

True Colors

Waukesha trademarked the stairway of black blocks in 1926 and painted the rest of its oil field units in bright, contrasting "oil field" orange so they would be easy to recognize on the job.

as the country song goes, they continued to "dance with the one that brung 'em," even if he was slow, stiff and didn't know the latest steps. Steam engines were difficult to transport over rough terrain, burned tons of coal that had to be hauled cross country and required gallons of water scarce in the arid oil region. To top it off, steam was a notoriously inefficient way to generate power. Internal combustion power plants were more convenient, powerful and efficient than steam units and they burned an oil field product, gasoline. The potential market was huge, first in the United States, then the world's largest producer of crude, but also in Venezuela, the Persian Gulf and East Asia, which some petroleum geologists claimed had more oil than Texas.

DeLong was named oilfield sales manager and set out to persuade American oil men to change dance partners. In Houston he met Louis M. Pearce, who told him that oil drillers needed an engine that was powerful, dependable and portable. DeLong replied that he had a shop full of people who could build Pearce any kind of engine he desired. In 1924, Pearce organized the Portable Rotary Rig Company and purchased a pair of Model EU engines from DeLong. It was Waukesha's first sale in the oil field and the start of a seventy-year relationship with the company that a decade later renamed itself Waukesha Sales & Service.

DeLong's experience with Pearce showed him that the way to make sales in the oil fields, and other markets as well, was to reach the person who actually purchased the equipment. In the oil market, it was the field supervisor or "farm boss" who bought the gear for his rigs, not the company executive. The farm boss didn't see the ads Waukesha placed in trade journals like *Oil and Gas Journal* and *Petroleum Engineer,* because he didn't read them. He did read his mail, though, so DeLong developed a direct mail program to reach the oil farm boss. The mailings pointed out that Waukesha's gasoline engines were not only as durable as steam units but superior. Writing and designing the mailings was easier than learning who should receive them. DeLong and his assistants traveled the roads and rails from the wetlands of Louisiana to the dry prairies of Oklahoma, west to the foot hills of the Rockies in New Mexico and beyond to California. The card file they compiled, and card file it was, listing the names and addresses of oil farm bosses, was an invaluable marketing tool. When the representative from Portable Rotary Rig or another distributor made his sales call, the farm boss recognized the Waukesha name. He had heard from the company. Direct marketing was not new in the 1920s, but it was not as widely used as it would be in the future. It was especially unusual for a manufacturer to target end purchasers. But it worked for Waukesha. By the time he moved up to plant manager in 1928, DeLong had established Waukesha Motor in the energy service industry and the company has been there ever since.

Oil drillers were not the only people who needed stationary power units in rural or remote locations. Cotton was still a principal crop in the southern states and most of the harvest was processed at small, locally-owned gins. Electric power had yet to come to most of rural America and less so in the

A Common Sight

Natural-gas-fired Model WOK engines powered the draw works and supply pump on this oil rig. By the time this photo was taken in the 1930s, Waukesha engines were a common sight in the American oil patch.

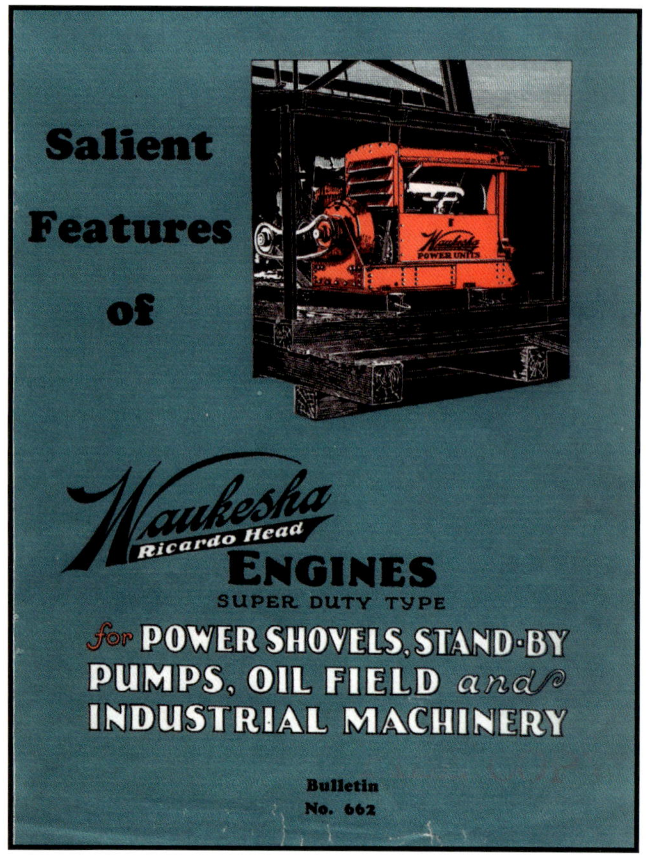

Diverse Markets

Product bulletins from the 1920s illustrate the diverse markets for Waukesha motors: agricultural and industrial machinery; construction equipment; pumps for oil, gas and water; fire fighting; as well as automotive transportation.

southern states than elsewhere so cotton gins ran on steam or water power. The same gasoline motors that ran oil rigs in Oklahoma, for example, could also run cotton gins in Mississippi, and Waukesha developed this niche as well. In the Pacific Northwest, loggers were at work in rugged mountainous terrain. Felling the old-growth timber with axe and crosscut saw was hard work. Moving the huge logs over mountain tops or down treacherous slopes was even harder and all but impossible for horses or the tractors and trucks of the day. Cable hoists powered by Waukesha's 1920's industrial models—BUX, DU, EU, XAH—could skid or loft the timber over hills and valleys to saw mills which could also be run by Waukeshas, or to roadside "landings" where they were loaded onto highway trucks.

As water well drilling and pumping equipment improved and agriculture spread in arid regions of the western states, Waukesha engines could run pumps that transformed desert sand into lettuce patches so Americans in northern states could eat fresh salads in January. In the mid-1920s, the irriga-

1924
The "turbulent" Ricardo Cylinder Head, the first Model X motors and the first six cylinder engines are introduced.

1925
Harry Horning is elected President of the Society of Automotive Engineers and President of Waukesha Motor.

1926
The initial offering of Waukesha stock is listed on the Chicago Stock Exchange at $25.

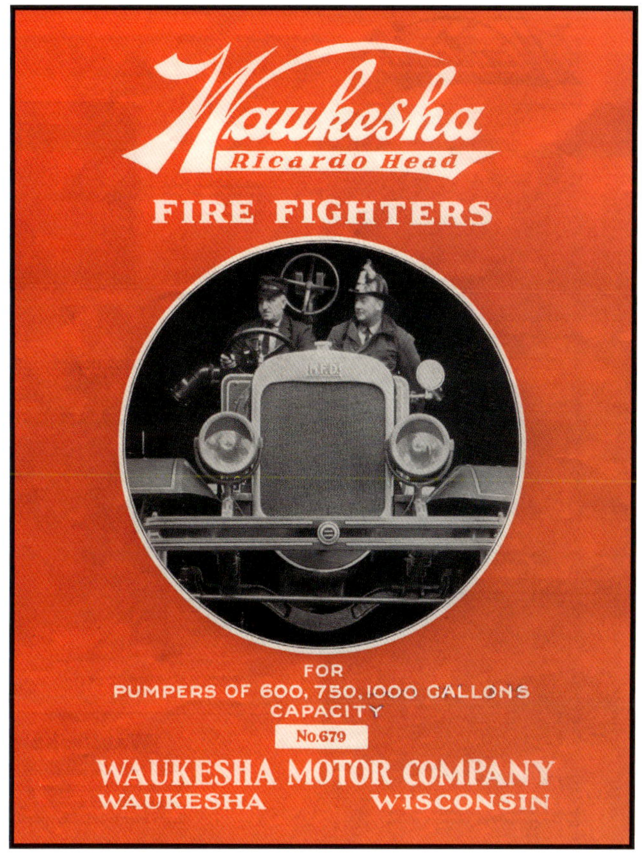

tion water pumping niche was as promising as oil's. Commercial aviation had also taken off and created new markets for motors, and not just in aircraft. Starting in 1926, the United States established a system of "airways" to safely guide air traffic from city to city. The air routes were marked by electric beacons that, in remote locations, ran on generators turned by gasoline engines. The Kohler Company was a leading supplier of aviation beacon systems and Kohler purchased many of its motors from Waukesha.

Over the course of the 1930s and certainly after World War II, as the electric power system expanded and as heavy-duty off-road transportation equipment improved, these niches diminished in the United States. In the 1920s, however, they represented an important market for Waukesha. Their decline in the 1930s spurred the company to develop markets overseas where conditions remained very similar to those of the United States in the 1920s. The term "Third World" was not coined until the 1960s, but the underdeveloped state it defines existed in the United States in the 1920s and persists through-

1927
For the first time, shareholders receive more than $1 million in dividends.

1928
J.I. Case orders 2,200 Model V motors for its new combine, helping to set a record for orders received in one year.

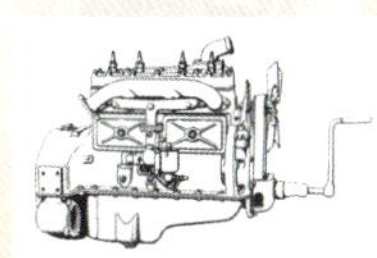

1929
The team of J.B. Fisher, Arthur Pope and Harold Wiles designs and builds the first Cooperative Fuel Research engine to be used to standardize grades of fuel worldwide. Company profits reach $2.2 million.

Model X

The Chicago Fire Department used a Kohler generator powered by a Waukesha Model XAK for its floodlight truck in the 1930s. Introduced in 1924, more Model X series engines were built than any other Waukesha model.

out the world today. Waukesha started producing engines for the "Third World" in the 1920s and continues to make them today.

The developed world also offered new markets for Waukesha motors. The United States was becoming an urban nation, with the 1920 census reporting for the first time that a majority of Americans lived in cities. Municipal governments had to expand services or, in the case of smaller cities, introduce them. Waukesha engines were installed in water pumping stations and sewage treatment plants. Fire departments were expanding and modernizing, with trucks replacing horses and gasoline pumps forcing the steam pumps into retirement. Likewise, gasoline-powered buses were replacing the electric-powered transportation systems that most cities had established twenty years earlier in the wake of "hydromania." Even with the development of the electric power grid, a market still existed for generator-engine sets in industry or for emergency back-up power in hospitals, for example. Finally, growing cities meant more construction work that required more construction equipment that stimulated another market where Waukesha was already established.

In 1924, Waukesha shipped nearly six-thousand-five-hundred motors, ten times as many as in 1914, with the BUX and the EU the top sellers. Many were sold as stationary power units, but construction-equipment manufacturers like Ingersoll-Rand and Harnischfeger also installed them in their products. That same year, Waukesha introduced its first six-cylinder motor, the

Model 6A. Not much more than an experimental model, the 6A did not register many sales, although Sterling Truck purchased a few. Improved versions of the 6A appeared over the next few years. In 1928, a 99-horsepower 6AB installed in a six wheel Fageol truck in California was hauling a load of sixty-four thousand gross weight over a two-hundred-and-fifty-five mile route every day. "As a daily diet this is putting any engine to a very strenuous test," was the opinion stated in *Turbulence*. Despite this promising start, the six cylinder engine did not come into its own at Waukesha until the late 1930s.

The development of a six cylinder motor was yet another confirmation of the company's commitment to research. The atmosphere of the place was described by Arthur Pope, who joined the engineering department in 1924. Many years later, Pope reminisced that he was "first attracted" to the company while reading Automotive Industries magazine. It "carried an article about a Waukesha truck engine which gave 100-ton miles per gallon. The illustration showed a modern looking Ricardo head engine and indicated that unusual things were being done at Waukesha. So I went to see and it was love at first sight." Pope was later named head of the CFR engine project and was thereby given an opportunity to make a contribution to engineering and to the world at large that few of his colleagues would be granted.

Of the man who gave him that opportunity Pope said, "The quality in Mr. Horning which stands out clearest…he was a thinker and a doer. There are many thinkers in the world and many men of action, but a combination of the two is rare." Horning's combination of gifts was coincidentally recognized in 1925. His fellow engineers had elected him President of the American

City Service

As American cities grew in the 1920s, municipal services like snow removal had to expand with them. Waukesha supplied the engines not just for power shovels and trucks, but for public water systems, sewage plants and other city services.

Society of Automotive Engineers for the 1925 term. At the same time the president's seat at Waukesha Motor was vacant, Conrad Haertal having resigned and moved up to chair of the board. The board of directors chose Horning to succeed Haertal and decided to announce the appointment on the very day he was handed the gavel as head of the SAE. They wired a copy of the resolution appointing him president to Horning while he was at the SAE convention. The thinker and the doer were honored together.

Horning then proceeded to demonstrate both sides of his personality. The doer introduced the XA series of engines, which stayed in production for nearly forty years. With one-hundred-ninety-one units shipped, Waukesha manufactured more of the XA series—which also included the XAH and XAK—than any other engine series. The thinker invited Harry Ricardo to address the Society of Automotive Engineers and invited him to tour the plant in Waukesha. Ricardo was only one in a long line of distinguished engineers to visit the company at Horning's request.

The doer was also a first-class business manager. In all but one year between 1923 and Horning's death in 1936, Waukesha Motor paid out six-figure dividends to its shareholders. In fiscal 1927, the shareholders received a whopping $1.2 million in cash and stock. Net profit after the dividend was $972,000. That same year the thinker, equipped with letters of introduction

Trackless Train

The Trackless Train was a promotional tool for the MGM movie studio. In 1925, it traveled from coast to coast to promote MGM's movies, stopping along the way in Waukesha. The "locomotive" was powered by a U-model Waukesha.

from colleagues at Ingersoll-Rand and from Wisconsin Governor Fred Zimmerman, toured motor manufacturing plants in France, Italy and Germany. He also journeyed to the laboratory of the Swedish *Civilingenior* Jonas E. Hesselman.

Hesselman had gained fame as an original thinker who had invented a motor that combined features of the diesel and the conventional combustion engine and could therefore burn a wide range of fuels with little loss of efficiency or power. The Hesselman engine fit into Waukeska's standard operating philosophy. No one in the United States was using the Swedish technology, which meant Waukesha could introduce it and have yet another new motor for another marketing niche. Because it could burn a wide range of fuels, the Hesselman was also ideally suited for "Third World" applications, where anything from high-octane gasoline to low-grade fuel oil might be poured into its tank. Horning's visit with Hesselman, like his earlier visit with Harry Ricardo, was fruitful. In a few years, Waukesha obtained American licenses for Hesselman engines and was marketing them worldwide.

It was Horning's personality—the doer and the thinker combined—that supplied the vision and the mission for Waukesha Motors throughout his lifetime and for many years after.

The Lone Eagle

Waukesha developed a close relationship with the Fageol Truck Company of Oakland, California in the 1920s. To promote the collaboration, "Drive-'em Jack" drove his Fageol from Oakland to Waukesha. Taking advantage of the publicity generated by the trans-Atlantic airplane flight of "Lone Eagle" Charles Lindbergh in 1927, and even though he never left the ground, "Drive' em Jack" became "The Lone Eagle of the Pacific."

THE ROARING DECADE

Model XAH

With sales in the tens of thousands, the Model XAH engines in the test room here helped make the 1920s roar at Waukesha.

It was the Roaring Twenties, so named for its extravagant popular culture and exuberant economy. The Twenties roared their loudest during the second term of President Calvin "Silent Cal" Coolidge, 1925-1929, and a good share of the noise emanated from Waukesha. It was one of the most productive and profitable periods in the company's history. At its start, the company had approximately one thousand workers turning out twenty different engine models, including the first Model V. In its own right, the 50-horsepower Model V was a solid seller. It was also the base model for the V series, which included the larger bore 1929 Model VK and the VI series of the 1930s, with four models designed for motive, industrial and marine applications.

Next, Waukesha introduced the XA series that, in only twelve years of production became the company's best-selling model. The majority of these motors went to International Harvestor for trucks and tractors but tens of thousands were sold for other applications. The litany of new motors continued in 1927 with the major addition of seven six-cylinder models based on the 1924 Model 6A. More sixes came down the line in 1928 and 1929. They represented Waukesha's first serious departure from the four-cylinder design.

INTERNATIONAL TRAIL

Dedicated to the Wider Development of Transportation Throughout the World.

Published Quarterly by the International Harvester Organization 606 South Michigan Avenue Chicago, Illinois

Vol. 5 *No. 4*

September, 1928

Out of Africa

The Model XA is one of the most important engines in the history of Waukesha Engine. The XA was the best-selling and most widely applied engine series of the 1920s and went a long way towards establishing Waukesha's reputation for rugged dependability. In 1928, a Model XA demonstrated its capabilities by powering an International Harvestor "Special Delivery" light truck for 6,718 miles through the mountains, rain forests and deserts of Africa.

Clyde King, an International sales representative, started the trek in Nairobi, now capital of the east African nation of Kenya. He motored west through the Kenyan highlands, then down into the equatorial forest of the Congo River valley, up to Lake Chad and on to the city of Kano in northern Nigeria. Averaging about two hundred miles a day, King took nineteen days to travel the 3,800 miles from Nairobi to Kano. The engine was well broken in by the time he arrived. In Kano, he met an adventurer named Sir Charles Markham who had a notion to return home to England by way of the Sahara Desert.

Markham purchased the XA-International, along with a few gasoline cans, a modicum of motor oil, and two—not four—new tires. Keeping the two old tires for spares, he set his sights on the caravan route that ran north from Kano for 2,818 miles over parched rock fields, dusty hills and searing sand dunes to Algiers on the Mediterranean Sea.

He was joined in the venture by a footloose Danish expatriate, the Baron Frederick Bror Blixen-Fineke. He had come to Kenya in 1914 with his wife Karen to establish a coffee plantation but soon came to prefer wandering in the wilds to staying put on the farm. Leaving Karen to tend the coffee trees, Blixen disappeared for months at a time. A truck ride across the desert was his cup of tea, so to speak. The route was a well-marked camel trade route and the two adventurers expected to find stores of gasoline and—more important—water along the way. Nonetheless, they got lost, missed an oasis and

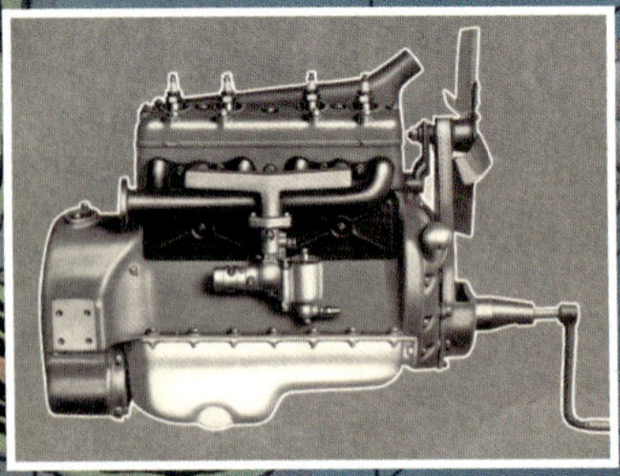

were dying of thirst when—in a scene that could have been borrowed from an African adventure yarn—they spotted a set of water drums shimmering on the horizon. With its radiator nearly bone dry, the XA carried Markham and Blixen to the life-saving water.

Although occasionally overheating in the one-hundred-twenty degree desert temperatures, the Waukesha XA kept the International's wheels turning for the sixteen day trip to Algiers. Despite extreme conditions the engine made fifteen miles to the gallon of gas and consumed hardly any oil. As a publicist for International stated, the motor ran through "unpeopled, unmapped, immeasurable waste spaces…under blistering desert heat; [with] water tank and radiator nearly empty, part of the way lost and the truck hub-deep in sand—and through it all, the tireless rhythmic beat of the motor that never misses, never complains."

In Algiers, Markham parted company with Blixen, but not from the XA-International. He loaded it on a boat bound for France, then drove it a few hundred miles from Marseille to the English Channel. He then ferried the truck to England, where it continued to run as well as it did in the desert.

As for Blixen, he would have faded from memory, but for his wife Karen. The coffee plantation failed in 1931 and she returned to Denmark. There she turned her African experience into works of literary art under the pen name Isak Dinesen. In 1985, a movie starring Meryl Streep and Robert Redford was made, loosely-based on one of Dinesen's books, *Out of Africa*. The XA-International truck did not get a part in the film.

International Trail
September, 1928

Across Africa by Motor Truck

3,800 Miles in 19 Driving Days

By Clyde N. King

FINDING myself in Nairobi in December, 1927, and desiring to reach Nigeria and the Gold Coast Colony before the rainy season started, I decided to drive through Kenya, Uganda, and Belgian Congo to Stanleyville and go by boat down the Congo River to Boma and up the coast to Nigeria.

It happened that I had met Father Thomas Gavan-Duffy on the British India boat coming down from India to Mombasa. Father Duffy intended to visit missions in Tanganyika, Kenya, Uganda, and the Sudan, and then leave Africa by way of the Nile boat from Rejaf to Cairo. He expressed the desire to make the trip with me and it was arranged that we were to meet at Kampala, on Lake Victoria Nyanza on December 20.

Father Duffy speaks French fluently, and the success of this trip I attribute largely to his knowledge of languages, travel, and his very pleasing personality.

Mr. King and his International Special Delivery truck at Buta, Belgian Congo, about two hundred and ten miles north of Stanleyville. It is common belief that the African elephant is untamable, but at the Buta mission, Mr. King found many of them pulling two-wheeled carts, plowing, and being used for other domestic tasks.

ful in loading 64 Imperial gallons (76.8 gallons) in cases.

At 5:30 A. M., December 17, I left Nairobi. The speedometer registered 22

feet) was uncomfortably cool. After the sun came up it grew warmer and by the middle of the morning it was very hot.

The road toward Naivasha is a gradual

INTERNATIONAL
TRAIL
African Adventures
of an
American Truck
INTERNATIONAL
Algiers
Sea
LAGHOUAT
GARDAIA
EL GOLEA
FORT MCMAHON
TIMIMOUN
DESERT
REGGAN
Tanezruft Desert
WATER CASKS
TABANKORT
IN TASSIT
ZINDER
NIGER RIVER
Kano
Lake Chad
FORT LAMY
BAMBARI
BANGASSOU
MONGA
DJAMBA
BUTA
FARADJE
WATSA
MAHAGI
BUNIA
IRUMU
STANLEYVILLE
BUTIABA
MASINDI
KAMPALA
JINJA
BUSIA
MUMIAS
KISUMU
Lagos
Lake Victoria

Model 6AL

Waukesha introduced its first six-cylinder engine in 1924 and the Model 6AL in 1925. The Milwaukee Fire Department used a Model 6AL on a ladder truck in 1927.

If that were not sufficiently innovative, the company produced the first model of the landmark Cooperative Fuel Research engine in January 1929. By then, Coolidge was packing to return to Vermont and Waukesha Motor had seventeen hundred men and women on the payroll manufacturing more than forty engine models.

More people, more production, required more plant. In the latter half of the 1920s, Waukesha purchased more frontage on St. Paul Avenue for additional office space, enclosed alleyways between buildings and then nearly doubled its shop space with the addition of Buildings Five and Six. To finance these additions, along with the machinery and material they would house, Waukesha made its first public offering of stock. In October 1926 one hundred thousand shares of common stock were listed on the Chicago Stock Exchange at $25 per share. The timing was right, as the stock market was heading into its late-1920s rise that saw the Dow-Jones industrial average increase from 140 in January 1927, to 375 in early October, 1929.

Also rising were profits and sales at Waukesha Motor. Net profits from operations approached the $1 million mark in fiscal 1927 ($10.8 million in 2006), hit $1.4 million in 1928 ($15.4 million) and jumped to $2.2 million in 1929 ($24.5 million). In November, 1928, *Turbulence* reported that during the previous October the "number and value of engines shipped...went over the

million dollar mark for the first time in the company's history." October 1928 was also the second month in company history—August was the first—that Waukesha's sales force booked more than one million dollars in new business. "Can this be kept up?" asked the editor, who then answered. "It's got to be."

In order to keep up sales, Advertising Manager P.C. Ritchie devised a new advertising campaign. Print was the only serious advertising medium for a company like Waukesha and the company made use of trade journals like *Bus Transportation, Oil Weekly, Pit and Quarry, Fire Engineering* and *The Timberman* to reach its diverse markets. Building on DeLong's success in the oil field, Waukesha also used direct mail to farmers, who were found to take greater interest in tractor motors than other end users and responded well to postcard mailers and clip-out response forms in magazines.

In Waukesha's early years, company spokesmen relied on technical information to sell motors. They believed that if a potential purchaser knew all about the superior qualities of Waukesha motors, he could not help but buy one or one hundred. Information was also important since gasoline motors were new. How they worked and what they could do had to be explained, not just to end consumers, but to manufacturers and distributors too. Flexibility was also part of the sales strategy and agents were to make it known that if Waukesha didn't already have a motor to fit the customer's needs, it could make one.

Horning had defined this approach as "science applied commercially in the development of a technical product." It was the standard operating philosophy of the Motor Company. The experimental labs, the constant stream of new motors, the interest in the latest developments in engine design, all stem from it. But technical information alone would not a motor sell.

Waukesha also had to pay attention to the advertising adage that says, "sell the sizzle not the steak." In the early 1920s, the company presented its

Gas Street Car

The Versare Company advertised its unusual bus as "America's First Gas-Propelled Street Car," to contrast it with the familiar electric-powered street railroads and trolley buses it was designed to replace. Mounted on pivoting carriages, both the front and rear wheel units could turn to help the bus clear tight corners and navigate narrow city streets. It was powered by a six-cylinder gasoline-burning Waukesha.

More Space Required

The cover of a promotional booklet describing the plant expansion of the late 1920s. Completed in 1929, the expansion nearly doubled the size of the plant.

Advertising manager C.P. Ritchie sought to make Waukesha a household word and designed print advertising extolling the engines themselves instead of their applications. He wanted to make the Waukesha engine "a national institution."

sizzle in the form of ads depicting Waukesha motors in action and the more dramatic the better. Hauling logs up mountainsides, powering trucks across deserts and fighting fires in crowded cities were popular themes. Of necessity, the ads also featured Waukesha's customers, who were happy to be in an ad they did not buy. This campaign was the victim of its own success. As the number of Waukesha users grew, the number of times each one could appear in an ad fell. If the purchasing agent for Peter Pirsch fire engines, for example, saw his equipment featured in one Waukesha ad, he also saw his competitors featured in ten others, and he did not smile. Rather than risk alienating all of its customers, Waukesha changed its ads.

The focus shifted directly to the motor. Ad copy elaborated on "The Why of Reinforced Bearing Caps," a "Deeper Crankcase for Longer Life," and "Center Bearings That Stand UP," all positioned next to appropriate graphics. An ad titled "Years Without Repair" for example, had an illustration of the pyramids of Egypt to reinforce the point. With this campaign Ritchie took a first step towards his goal of making the Waukesha name a nationally recognized brand. If Wrigley's meant chewing gum and Ford meant cars to the American consciousness, than Waukesha should mean engines—and not just to purchasing agents, oil farm bosses or engineers, but to the "average man on the street."

He proposed a pricey series of ads to run in general interest magazines such as *Colliers, The Literary Digest* and the *Saturday Evening Post.* For advertisers in the 1920s these large circulation magazines were the equiva-

Men At Work

Workers assembled oil field units in early 1929, when the work force numbered seventeen hundred.

lent of television in the 1990s. "It must not be expected therefore that our ads should sell goods," Ritchie said, "but merely that they should create that good will that would make a call from our salesman an agreeable incident in any place he might go. They are his introduction and should give him the prestige and standing that our product and our company policy justify. National advertising on a consistent and continuous plan is the only economical means to attain this position—a national institution." Ritchie was not the first nor the last to reach for such status. Xerox did it in the 1960s, but a better comparison is the computer chip maker Intel, whose "Intel Inside" stickers appear on computers made by any number of manufacturers. Ritchie wanted to see the equivalent of "Waukesha Inside" on trucks, cranes, compressors, oil rigs, tractors and electric generating units no matter who assembled them and wanted consumers to know it was there.

"Our responsibilities now are greater than ever for we have twice as big a shop to fill and it is only by our efforts that the wheels can be kept turning at full capacity. I predict however that if we seize the opportunities that are before us...we will have no difficulty in maintaining the pace and increasing it so that a year from now we will be putting up two more big buildings as we did this last year." Ritchie wrote those optimistic words in late 1929. He was not yet aware of the gravity of the economic crisis the entire industrialized world was about to face.

He was not alone. At the annual stockholders meeting in mid-October 1929, Horning announced a "new financial plan." The offering of 1926 had been so successful that Waukesha "stock has reached a price that is out of the trading zone for securities of our type." Therefore, the company could meet its future needs for capital "through the sale of securities" and still pay "substantially larger dividends." The company would issue five hundred thousand new shares of stock. One thousand would be sold, one thousand held in reserve and the shareholders would receive a three-to-one split with the remainder. To no one's surprise, the shareholders enthusiastically voted in favor of the plan.

Horning also announced that the company had completed a four-year program to develop an automotive diesel engine. The prototype was running and on display in the yard. The capital raised from the new stock sale would be used "for the further exploiting of this work," and for studies of "new methods of burning heavy fuels." Waukesha would also begin research on "an aviation diesel engine."

About a week later, the Dow began the slide that would see it lose one-third of its value by the end of the year. The decline continued until the summer of 1932 when the Dow bottomed out in the low forties. Shocked by the October 1929 crash, Waukesha's directors sent a special letter to the stockholders on November 15 asking them to rescind their approval of the new financial plan. Equally shocked, the stockholders agreed. Ritchie's "national institution" advertising program was indefinitely postponed, as was work on the diesel aircraft engine. The Great Depression had begun.

First Diesel

Four years in development, Waukesha's first diesel engine was placed on display for the stockholders meeting in mid-October 1929. Two weeks later, the stock market crashed, signaling the end of the 1920s boom and the start of the Great Depression. Diesel development all but halted until the end of World War II.

Cooperative Fuel Research

Howard Wiles tested the CFR engine that he and Arthur Pope assembled in time to meet the deadline of the Society of Automotive Engineers in 1929. As the standard instrument for measuring the quality of gasoline sold worldwide, the CFR has had an impact greater than any other Waukesha engine made.

THE CFR® ENGINE

A motorist drives up to a gasoline pump and gets out of the car to purchase fuel. She taps the keypad to choose the payment mode then hits the button near the nozzle to select the type and grade of gas to put in the tank. Her decision is based on the manufacturer's recommendations for her car and on her own family budget. She does not have to worry about the quality of the gas. If her car runs best on high-octane premium grade, she can tap the button above the nozzle labeled premium and feel confident that premium is what she's pumping. Cheating occurs, of course, but if the fuel delivery system from refinery to fuel tank is operating by its own standards of accuracy and integrity, the premium pump always delivers premium gas.

Duplicate the scenario many times over from the 1930s to today. Edit it to include semi-tractor drivers, work and pleasure boat owners, aircraft engine builders, industrial power plant workers, all the myriad users of gasoline or diesel oil, who safely presume that the fuel they pump into their tanks will perform as expected no matter when and where they buy it.

It was not always so. In the opening decades of the petroleum century, variations in the quality of fuels vexed the entire gasoline motor community. No one was more troubled and had more reason to be troubled than engine manufacturers. They could design and build a motor that ran at full power and efficiency on the fuel delivered to the factory, but coughed and misfired when fueled one thousand miles away. Oil refiners were also justifiably concerned. They wanted to deliver a product that met basic standards for "combustion suitability" around the world. In other words, they wanted premium gas in Hong Kong to be the same as premium in Kansas.

Landmark

Howard Wiles and the original CFR engine at the ceremony identifying it as an International Historic Mechanical Engineering Landmark in 1980.

In the 1920s, the engine builders and the oil refiners formed the Cooperative Fuel Research committee to develop standards for fuel and a means to measure them. Following the lead of Harry Ricardo, they accepted the volume of octane in a fuel mix as the gauge of its suitability for combustion. They also determined that the best way to measure octane was by means of a "knock test." Knock is defined as the "audible sound of undesirable detonation of a fuel." The more knock, the more "undesirable" the fuel. To measure knock they needed a specially designed motor.

Harry Horning and Waukesha Motor had been working on the fuel problem for many years. He was a prominent member of the Cooperative Fuel Research committee, so was the man he hired in 1924 to be the company's chief research engineer, Arthur Pope. By 1928, as Pope recalled, the CFR committee had decided, after much debate, "that a standardized single-cylinder engine must be produced as the first step in developing a gasoline knock test." The group then deadlocked on "the issue of L-head versus variable compression valve type." Pope said he resolved the issue because he "had the good fortune of dreaming up a very simple gear linkage which permitted interchangeable L-head and variable compression overhead valve cylinders being installed on the same basic crankcase."

In December 1928, Horning presented Pope's idea to the committee. They accepted it and suggested that Waukesha Motor build a prototype for exhibit at the next annual meeting of the Society of Automotive Engineers the following month. Horning said yes and passed the order down to his chief engineer, James. B. Fisher, who shared it with Pope, who called for Howard Wiles.

Wiles had walked into the personnel office a few months earlier, fresh out of Iowa State University, with his senior thesis on the design and construction of a variable compression single cylinder engine under his arm. He was hired and soon assigned to the CFR project. He went to work fashioning many of the parts for the engine, including the crankshaft, on a bench lathe in the experimental lab. After many a long day and night—forty-five, to be exact—he and Pope completed the engine in time for the SAE meeting. The CFR committee accepted it and the Waukesha motor became the instrument on which engineers conducted the standardized test for measuring octane and thereby standardizing the grades of fuel throughout the world. The engine was ready in 1929, but the octane test was not perfected until 1931 and that is when the Waukesha CFR went into production.

The company retained possession of Wiles' 1929 engine and the first production model. Texaco and Standard Oil bought the next four, followed by nearly every other American oil refiner, then by General Motors and Ford, the U.S. Bureau of Standards and the Massachusetts Institute of Technology, to name a few. In short time, Waukesha CFRs were shipped to England, where they replaced Harry Ricardo's test engine, then to Mexico, Argentina, Germany, Poland, Romania, the Soviet Union, Sumatra, Japan and points in between. In the seven decades since production began, Waukesha has produced well over 5,000 CFRs.

Although a bit surprised at their engine's initial success, Waukesha's management soon realized its significance and acted to maintain its corner on the test market. Within the company, the CFR department became a separate and elite unit. "Building these engines entailed a great responsibility," wrote James DeLong in a booklet commemorating the twenty-fifth anniversary of the first CFR. Ordinary line-production methods were not suitable for these precision instruments. Not only was it necessary to set up manufacturing and inspection standards specially suited to these engines, but it was desirable to departmentalize and isolate their production." To set a standard for the world, special measures had to be taken and maintained for decades. Although improvements and additions have been made over the years, the CFR has remained in production longer than any Waukesha engine.

The CFR was the first of a series of fuel test engines the company would design and build. In 1931 the variable compression ratio model was modified to study the ignition quality of diesel fuel. In the late 1930s and during World War II Waukesha also developed the Counter Weighted Aviation engine for testing fuels at high speed, the Cooperative Oil Test engine for testing lubricating oils and the Cooperative Universal Engine for testing aviation fuel. The term Cooperative in the title points out that the engine was the product of research and design performed by competitors in both the engine-manufacturing and oil-refining industries who agreed to work together to meet a common goal. The spirit of Harry Horning lived on and was embodied in the cooperative ventures.

In 1980, the original CFR engine was recognized as an "International Historic Mechanical Engineering Landmark" by the American Society of Mechanical Engineers. Howard Wiles had watched over his engine and, when he retired in the 1960s, took it home with him. He agreed to return it to the company so it would remain on permanent display as the "Landmark" it is. Wiles' care, Arthur Pope's pride and the devotion the CFR inspired in the other people who worked on it over the years were well justified.

Waukesha has produced well over one million motors in its first century. None has had a greater impact on the quality of life of billions of people around the world than the CFR.

CUE

Workers assembling the Cooperative Universal Engine to measure the quality of aviation fuel, that Waukesha introduced in 1938.

Hard At Work

Shop workers on the Hesselman line during the Great Depression. In 1933, Waukesha recorded the first loss in its history and reduced employment from 1,700 to 700 workers.

CHAPTER TWO

DEPRESSION AND WAR

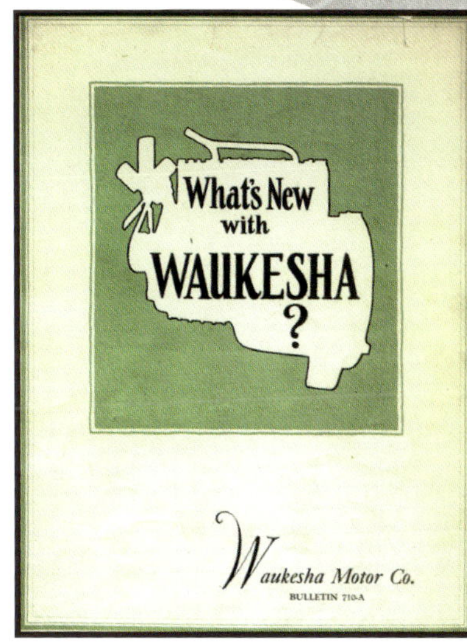

WHEN HARRY HORNING AND SAM PERKINS CAME TO WORK ON THE MORNING OF APRIL 6, 1931, THEY FOUND THEIR DESKS COVERED WITH BOUQUETS OF FLOWERS SENT TO THEM IN HONOR OF THE COMPANY'S TWENTY-FIFTH anniversary. The ceremony to mark the birthday, if ceremony it can be called, consisted of friends and co-workers visiting the two founders and shaking their hands. In addition, James Remington, who left the company in 1919, was called upon to reminisce about Waukesha's earliest days for the readers of the newsletter *Turbulence*.

He recalled how young Harry Horning rode a bicycle to and from work, how the daughter of company president Allan Stebbins would call the garage every day and, with a giggle in her voice, ask "Hello Jimmy, is this the motor works?" and how, as low man on the corporate totem pole, Remington had to come in early, stoke the boiler, dust the desks, sweep the shop, and swab out Sam Perkins' cuspidor. He also carried car batteries across town to be charged at Metal Structures and delivered drawings of new motors to the city engineer for blueprints. When a customer wanted to visit the "factory," Horning and Perkins did their best to schedule the tour after working hours. They would then collect as many coveralls and jackets as they could find and spread them at what appeared to be work stations around the shop so the sales prospect would think Waukesha Motor had not three or four but thirty people at work.

Remington's lighthearted recollection of Waukesha's past was in decided contrast to its grim present. Along with the rest of the industrialized world, the Motor Company was facing the most serious economic crisis it had known and it would take more than artfully placed coveralls to solve it. As always in busi-

Great Six

What's New with Waukesha, a bulletin published in 1929, featured "New Engines, New Facilities and a New Plant," including the company's most powerful engine to date, the 300-horsepower, Model 6OK, depicted here while mounted on the tilt test.

Comets

The Comet and Silver Comet engines were two of the hundreds of models Waukesha introduced to stay afloat during the Great Depression. About two hundred of the six-cylinder diesels were built between 1934 and 1936.

A generator unit is loaded onto Waukesha's heavy duty Sterling truck, 1934. The truck carried advertising for the Silver Comet but had a smaller 6D-125 Comet under its hood.

ness, the numbers tell the story. At the end of the fiscal year in July 1929, Waukesha's net income exceeded $1.8 million ($20.6 million in 2006). At the end of fiscal 1930, net income was $450,000. Income continued to drop in 1931, recovered to $485,000 in 1932, then plunged in 1933. For the first time in company history the ledger of Waukesha Motor recorded a net loss, with $123,000 in red ink ($1.8 million in 2006).

The company's performance mirrored that of the entire economy, particularly the automotive industry. The entire sector had contracted nearly two-thirds between 1929 and 1933, and all of the Big Three American automakers recorded heavy losses. Total wages paid were one-third that of 1929 and nearly one of every two autoworkers was unemployed. In April 1933, Horning reported that Waukesha Motor was running at about twenty percent capacity, "which was very good for the present times."

The number of employees had fallen from seventeen hundred to seven hundred. Each morning, machinists, laborers, skilled and unskilled, gathered at the plant gate waiting for a foreman to appear and call them into work. Few were called. Office workers fared a bit better. They were all laid off for two weeks without pay and everyone, from President Horning down to the receptionists, also took a ten percent pay cut. It was small comfort that Sam Perkins, the man who found the investors to take the company out of the garage that "Jimmy" Remington recalled, did not live to see it register its first loss. He died in August 1932 at sixty-nine years of age.

Not all the news was bad. The inauguration of Franklin Delano Roosevelt in March 1933 brought hope that a change in Washington might jolt the econ-

omy to life. Horning expressed his willingness to cooperate with New Deal programs to revive the system but no matter who held office in Washington, Waukesha had its own resources. The manufacture and sale of replacement parts and motors brought a steady stream of income. The diversity of the markets it served meant that Waukesha could expect at least a few orders for new motors from each of many sources. To be sure, nearly all sectors were down—orders from Waukesha's key indicator client, Ingersoll-Rand, fell from 3186 in 1929 to 243 in 1932—but none were completely out. Of all the sectors Waukesha served, the oil industry had declined the least. Production had fallen, but only by about six percent, a fraction of the declines experienced by other sectors. Many of its people would suffer genuine hardship but, with a little bit of business from here, a little bit from there, Waukesha Motor could make it until conditions improved.

The company was not in debt and, in fact, was acting as court-appointed receiver for the bankrupt Fageol Truck Company in California. It had also acquired a large share of Sterling Truck in lieu of cash for motors delivered. These were real, if dormant, assets. Waukesha also had a large cash reserve, which was both good and bad. Having money in the bank was better than not, but money in the bank was also money not at work buying materials and paying people to build motors. A strong cash reserve was a sign of underutilized capacity. Waukesha had the plant, the capital and the work force. It needed orders for motors.

continued on page 66

Diesels On The Move

The Silver Comet truck at the Los Angeles distribution center, with Comet and Silver Comet diesels on display.

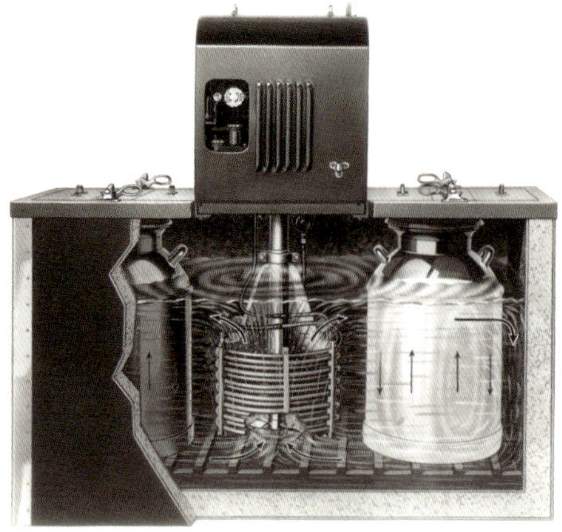

The Ice Engine

To shake itself out of the depression doldrums, the Motor Company attempted to move into the field of refrigeration. Since electrical service had yet to be extended to much of rural America, a market existed for on-farm cooling plants, especially on dairy farms. The farm experiment failed, but led the way to the railroad refrigeration and air conditioning markets, where Waukesha excelled for three decades. After railroad air conditioners came the Enginator electrical power plants, also for the railroad. From there it was an easy step to the much larger generator units that constitute a major market for Waukesha today. The 1000kw APG™ engine of 2006 has it roots in the Ice Engine of 1934.

The Ice Engine research team (below) visiting a dairy farm, 1934. The trailer contained demonstration models of the milk cooler and other cooling equipment.

The milk cooler and tank (above). The Waukesha home ice maker. It froze water into twelve-pound blocks which the homemaker then transferred to her ice box.

The MOST *practical* DEVELOPMENT SINCE THE INTRODUCTION OF ARTIFICIAL REFRIGERATION

THE WAUKESHA ICE ENGINE

• During the past decade, millions of homes throughout America have found new convenience, new economy and new delight in the use of mechanical refrigeration. Homes without electricity have been denied the joy of this new, modern development.

Now a gasoline ice engine has been developed that supplies power for a modern mechanical refrigerator, for a portable ice maker, for a milk cooler that cools milk quickly and cheaply. Now homes without electricity may have all these modern conveniences, and they really pay for themselves in the savings effected.

This new, unusually practical, modern development consists of a refrigerating compressor, and an air-cooled, four-cycle gasoline engine, all in one compact, precision-built unit. It is specially designed for the purpose, and perfectly balanced. It runs smoothly, quietly, and is so simple that a woman can operate it. This unique Ice Engine is the product of the Waukesha Motor Company, the world's largest builders of heavy-duty gasoline engines for agricultural and industrial purposes—a company whose reputation for engineering skill has won world-wide recognition, whose reliability and responsibility are of the highest.

The simplicity of the Waukesha Ice Engine is remarkable. The engine is automatically oiled by a simple splash system. The compressor is sealed and oiled for life. Fan blades in the flywheel throw a powerful air stream which cools the motor, compressor, condenser, and actuates the governor. The flywheel also houses the magneto.

The refrigerant used is methyl chloride, commonly used for heavy-duty commercial work, and chosen here because it is more efficient than other refrigerants generally used for domestic service. It has no disagreeable odor—is non-irritating, safe.

The engine starts with a handy ratchet lever—simple, sure and safe. Meets all the requirements of safety against fire hazards.

There are no joints in the gasoline supply tank below the gasoline high level. The safety fuel system delivers gasoline only when the engine is running; no fuel stands in the carburetor. Insurance risks are not increased—there is no open flame.

The Waukesha Milk Cooler and Waukesha Ice Maker may also be powered with the Waukesha Electric Ice Motor. However, the Refrigerator is powered only with the gasoline Ice Engine.

TURN THE PAGE AND READ ABOUT THESE MODERN CONVENIENCES

THE REFRIGERATOR

THE ICE MAKER

THE ELECTRIC ICE MOTOR

THE MILK COOLER

Cool Characters

James DeLong (left) with inventor Louis Melcher in front of an air-conditioned Fageol bus, 1938. Melcher brought his patents for cooling equipment to Waukesha in 1933 and stayed to manage the company's diverse ventures into refrigeration and air conditioning.

To get orders, the company turned to its strengths—research, innovation, diversification. Horning spoke of the "necessity for diversifying production…to secure a revival of its business under existing conditions." He cited the CFR, which was a consistent if low-volume seller, as one example of diversity. He also talked about the Hesselman "oil engines", the Ricardo Comet diesel, and the "Hy-Power" engines then in development. Venturing into the realm of the far-fetched but not impossible, Horning also announced that Waukesha was building its first steam engine and developing a gasoline-powered refrigeration unit for use in homes and on dairy farms located off the electric power grid. The farm "ice-cooler," as it was originally called, was reminiscent of the barn stanchions for milk cows that the company had briefly manufactured during the recession of 1921, and about as successful, but it showed Waukesha's willingness to try something new.

The ice cooler was powered by the new Waukesha Ice Engine, a one-cylinder, air-cooled gasoline motor that also ran a home refrigerator, ice maker and cooling chest. Waukesha's first and only entry in the home appliance market, the ice engine was a hot item in 1933. The company consulted with an advertising firm whose agents pointed out that Waukesha would be competing with established appliance dealers like Norge and Maytag. They also wondered if Harry Horning was enough of a corporate celebrity for Waukesha to use his name and fame to promote the cooler, thereby anticipating the custom of fashioning the CEO into a media star common in the latter decades of the century. A campaign to place Harry Horning's honest but homely visage on billboards across America was not launched, but Waukesha did outfit a special "exhibition" truck and trailer to tour the countryside to demonstrate the cooler to wholesalers, dealers and the occasional farmer.

The Refrigeration Division was created and sited in the former Jiffy-Jell gelatin plant in Waukesha. It produced the home refrigerator, cooling chest, ice maker and milk cooler, and promoted the Ice Engine as "The Most Practical Development Since the Introduction of Artificial Refrigeration." It was not practical enough and Waukesha discontinued the refrigeration division after about a year. Had the farm and home ice cooler taken off, it would have not have flown for more than a few years. By the end of the 1930s, Roosevelt's rural electrification program had brought central station power to all but the most remote of rural areas and pushed stand-alone equipment out of the mainstream market.

Waukesha's venture into refrigeration was not a complete loss, quite the contrary. In the mid-thirties, the nation's tottering railroads thought they

WAUKESHA CHRONOLOGY

1930
Evidence of the economic depression begins to appear: net income declines 75% from the previous year.

1931
Waukesha introduces its first production diesel engine, a Model DWF.

1932
Design work begins on Waukesha's one and only steam engine.

could lift themselves out of the Depression by recovering the passenger service they had lost to the automobile. Comfort, speed and streamlined design became their watchwords as they introduced luxurious new passenger cars and powerful locomotives to pull them. The Chicago, Burlington and Quincy Railroad inaugurated its high-speed diesel electric service out of Chicago in 1933. The trains were called Zephyrs because they ran as fast as the west wind. The Milwaukee Road introduced the Hiawathas, with their Brooks Stevens-designed cars, including the first "Vista-dome" observation cars. The North Western brought out its *400* trains, so named because they promised to carry passengers from downtown Chicago to downtown St. Paul in four hundred minutes—faster than any auto obeying the speed limits of the 1930s or the 2000s. All three of these trains, and others as well, approached speeds of ninety miles per hour on parts of their runs.

Opening windows for ventilation on high speed trains was not practical, nor did the dust and smell of the rail yard meld with the air of luxury the railroads hoped to present. They developed a logical interest in the new technology

Cooling Unit

The cooling unit assembly line, which produced less than two hundred units in its one year of life.

1933

Operating at only 20% of capacity, the company records its first net loss.

1934

The Refrigeration Division is organized to produce cooling units for rural homes and dairy farms.

1935

The Refrigeration Division is converted to the Railway Division to produce air conditioning units for luxury passenger trains.

FC Engine

A Model FC motor coupled onto an Ingersoll-Rand compressor to make a railway Ice Engine Unit.

Executives (right) inspecting an Ice Engine installed on a North Western parlor car.

of air conditioning. The North Western approached Waukesha Motor and, more than any other railroad at the time, worked with the company to develop air conditioning units that were self-powered and did not run off the locomotive engine. Building on its experience with the farm ice cooler, Waukesha developed engine/air conditioning units powered by the 1935 Model FC that could be installed on the undercarriage of individual rail cars. Starting with the North Western *400s* in 1935 and continuing onto the *Rock Island Rocket,* the *City of New Orleans,* and other new trains, Waukesha sold railway air conditioning units to over thirty railroads in the Midwest and western states, as well as Canada and Mexico.

Spinoffs have a way of continuing to spin off and Waukesha's venture into passenger car air conditioning led to the development of gasoline and LP-gas-powered units to refrigerate freight cars and generate electricity. By 1939, Waukesha was the number one manufacturer of air conditioning and electrical generating units for railroads, accounting for eighty-four percent of all units sold. The railroad experience also prompted an excursion into the field of automotive, commercial, industrial and domestic cooling in the second half of the 1930s. While the Model FC was the workhorse of railroad and industrial refrigeration and electrical generation, the 15-horsepower Model ICK—the smallest four cylinder Waukesha engine—became the leading power plant for refrigerated trucks.

Spinoffs showed promise, but Waukesha was a motive and industrial engine maker first and foremost. The commitment to producing ever more

The 400

The North Western Railroad's high-speed *400* train leaving Chicago, 1940. It was the *400* because it promised to carry passengers from Chicago to St. Paul, Minnesota in four hundred minutes. In 1935, the *400* became the first luxury passenger train with cars cooled by Waukesha Ice Engines.

Map of the railroads (left) using Waukesha Ice Engines in 1937, only two years after their introduction.

Passengers (below) enjoying air conditioned comfort in the lounge of the *Rock Island Rocket,* 1938.

powerful and efficient motors for diverse applications did not wane. In fact, the most dismal years of the Depression, 1931-'35, were among the company's most prolific. New models came out in a dizzying alphabetic display. The list of four-cylinder models included the DWL and DWF, Waukesha's first diesels; the VIS, VIL; and VIK, marketed as the "Four-Forty...For Combines, Harvesters, Tractors" and also the first models with overhead valves; the CH and CHK, "Four-Fifty...For Excavators, Shovels, Tractors, Road Building, Agricultural and Industrial Machinery"; the CSR and DKR, "Heavy Duty Fours"; the WOK, "Giant Four...For Cotton Gin, Oil Field and Industrial Machinery"; the 4-95, first with an F-head and one of the few models identified solely by numbers; and the versatile Model FC, which stayed in production for nearly forty years and over one-hundred-ten-thousand in sales. By the time Waukesha celebrated its fiftieth anniversary in 1956, approximately one of every seven engines manufactured was an FC.

The six-cylinder line was invigorated by the 420-horsepower 6LR0, Waukesha's first big-bore, big-stroke six. More significant for its engineering than its sales, the 6LRO design was adapted for diesel fuel and became the standard model for oil field and gas compression applications in the years after World War II. Introduced in 1930, it remained in production for thirty-seven years.

The number of six-cylinder models increased dramatically in the 1930s. Some of the entries in the catalog included the 60-horsepower 6ZK and 6ZKA, marketed as the "Courier Six...a reliable and sturdy six cylinder in compact form"; the 6TS and 6TL "Dispatch Six...For Motor Coaches, Trucks, Pumps, Electric and Industrial Machinery"; the 70-85 horsepower 6MS, 6ML, 6MK,

Cooling Off

Waukesha was proud of the engines on display at the Tulsa Oil Show in 1936, but stole the show with its air-conditioned hospitality room where prospective customers could find relief from the hot Oklahoma sun.

Spin-offs

The Interstate Electric Co. of Shreveport, Louisiana (left), kept its eleven-story headquarters cool with an air conditioning unit powered by a natural-gas-burning Waukesha Model 6BK, 1937. Interstate was one of many utilities in southern and western states that invested in cooling equipment demonstrations to increase sales of natural gas.

A Model ICK cooling unit (below) keeping the trailer cold for a Wisconsin milk hauler, 1940.

Horsepower

The first Waukesha promotional horse mimicked a 1920's pull toy and told customers *"You need real horsepower,"* not *"paper horsepower."* The toy horse was on display with engines at the 1931 Detroit Road Show.

The high-stepping Waukesha work horse (below) was featured as the company logo at the Chicago Century of Progress World's Fair in 1933.

6MZ "Express Six...suited to fast ten-twenty passenger coaches, 1.5 to 3 ton speed wagons and industrial duty;" the improved 110 horsepower 1938 Models 6MK and 6MZ "Express Super Six...for 2-4 ton trucks;" the 6HB and 6AB, "Highway Six...with instantaneous pick-up from idling;" the 90-110 horsepower Models 6SRS, 6SRL, 6SRK, "Transport Six;" the 6RB and 6RBR, 150 horsepower "Big Six;" and biggest of them all, the 250-300 horsepower Models 6OS and 6OK "Great Six."

The first six-cylinder diesels, Models 6D100, 6D125, 6D140, came out in 1934, but Waukesha manufactured fewer than two hundred diesel sixes in the 1930s. The emphasis in the six cylinder line was on gasoline models and Hesselman "oil engines." A gasoline-burning four-cylinder Model VI was modified into the 69-horsepower Model VIH in 1932 and nine more four-cylinder Hesselmans came out by 1935. The first six-cylinder Hesselman, the 171-horsepower 6ELH, came out in 1934 and new models—twelve in all—made their debuts before the end of the decade. The 1935 85-horsepower 6BKH, dubbed the "Marathon Six," was the most popular model of the decade and stayed in production until 1943.

The virtue of the Hesselman resided in its ability to run on a variety of fuels without burning up or losing power. "Practically any clean, refined high-speed diesel fuel," domestic furnace oil, kerosene or "other distillates" would do. In a pinch, the Hesselman could also burn a mix of gasoline and lubricating oil. The Hesselman was also economical, as a 1937 sales brochure pointed out. To produce one hundred horsepower for one hour, a Waukesha gasoline engine needed $1.22 worth of fuel priced at ten cents per gallon. A natural gas model needed only twenty-one cents per one hundred horsepower with gas priced at

Model 6LRH

Four Model 6LRH engines running oil field mud pumps and drawworks, 1942. Introduced in 1930, the 400-horsepower Model 6LRO was a prototype of the oil field and gas compression engines of the post-World War II era.

A race car (below) built by the Hemphill Diesel Engineering School, Chicago, and powered by a Waukesha Model 6D-140 engine set a world speed record of 125.69 mph in 1935. The Model 6D-140 was one of the few six-cylinder diesels Waukesha built in the 1930s.

ten cents per thousand cubic feet. A Hesselman burning three-cent-a-gallon fuel oil needed thirty-five cents worth to make one hundred horsepower per hour. Natural gas was more economical, but unavailable in many places. In most applications, the Hesselman was a better buy than the gasoline burner.

V-12 Cadillac

A one-of-a-kind V-12 Hesselman installed in a Cadillac coupe attracted a lot of attention at the Motor Company in 1938.

Ingersoll-Rand liked the Hesselman's versatility and economy. It purchased nearly two thousand of the 1930s models, teamed them up with its compressors and marketed them worldwide, particularly in underdeveloped locations. Fageol and Sterling also conducted experiments with Hesselmans in their trucks, but neither company was in a position to make a major investment in these engines or any other.

By 1935, Waukesha had one-hundred-twenty engine models in its catalog and even Harry Horning, who rarely saw a new engine design he did not want to build, thought that the research and development engineers should take a breather. He recommended that the company discontinue "fifty odd models." Horning had decided that he needed a respite too. The company had climbed out of the red and had at least returned to its 1932 level of income but was still struggling. Horning and his top managers had been examining operations

1936

Founder and President Harry L. Horning dies; he is succeeded by James DeLong.

1937

Economy improves; net income reaches $1 million and employment rises to 1,852.

1938

The "Little Depression" hits; Waukesha builds the Model 150 for Crosley autos and the first CUE test engine.

Spark Diesel

Ad for the Spanish-speaking market touting the Hesselman conversion for Chevrolet trucks, 1938. Waukesha acquired rights to Hesselman "all-fuel" technology and put a lot of engineering and marketing muscle into the line.

The Power Service Company, Waukesha's distributor in Denver, Colorado, installed a Model 6BKH engine in a truck for a long-distance hauler in 1940 and attached a "Waukesha Spark Diesel" nameplate to the hood.

and hired a consulting firm to review administrative and manufacturing practices. "Plant operations had grown up with the company as a result of its own development rather than as being compared with scientific research and developments in other plants," he concluded, "and that the recent rapid development of business affairs had shown the inefficiency of the present method of operation."

Except for the recessionary hiccup in 1921, Waukesha had chugged along successfully on an ad hoc basis, doing what was necessary to meet market demands as they arose. Unwelcome as it was, the Depression had forced Horning to examine his creation and see how he might improve its operation. Earlier in the year, after working with a committee headed by plant manager James DeLong and including veteran shop foreman Tony Herzberg, Horning had authorized a return to the piece rate system of compensation. The shop had

1939
After a month-long strike, Machinists Local 1377 and Waukesha reach agreement.

1940
United States begins multi-billion-dollar defense buildup; Waukesha introduces Series 145 and 160.

1941
United States enters World War II; Waukesha begins all-out effort for victory.

been operating on a "group rate" system that seemed fine in theory but proved to be less productive than the piece rate based on individual performance.

He was also reviewing purchasing, cash flow and accounting procedures to reduce turnover time. Its strong cash assets had enabled the company to weather the worst storms of the early 1930s and Horning wanted to increase that strength by reducing the amount of time between the arrival of materials at the receiving dock and the exit of finished motors from shipping. "It was better to have the inventory at a lower figure and a large cash balance, than to have so much money invested in materials and supplies." Horning was not about to invent "just-in-time" management, but his thoughts were headed in that direction.

Model WAKH

The Power Service Company was also ready when a customer requested a replacement engine for his RD-8 Caterpillar tractor. Power Service installed a Waukesha Model WAKH.

He was also confident that the economy and his company would grow again. To be ready for it, he proposed that—after a six-year postponement—Waukesha Motor Company stock should be listed on the New York Stock Exchange. One hundred thousand shares of no par value stock would be converted to four hundred thousand shares of capital stock valued at five dollars per share—one-fifth the value of the 1929 offering. The offering was accepted and the stock went on sale in November 1935.

The NYSE had yet to inaugurate the custom of inviting officers and managers to ring the bell at the start of the day their offering opened for sale. Had he been invited to ring the bell, Horning would not have been able to attend. After the annual meeting in October, he had checked into a hospital in Battle Creek, Michigan, for long term care of a "complication of ailments." He lingered in failing health past Thanksgiving and Christmas, and died in the early hours of January 4, 1936. He was fifty-five years old.

January 4 was a Saturday and so nearly everyone at the company knew Horning had died when they came to work on the following Monday. Engineering department head J.B. Fisher, a colleague and friend for twenty years, told his fellow board members that Horning's death was not only a personal loss to everyone at Waukesha, it was also a blow to the company. He said that "every man should put forth extra effort…in fidelity to Mr. Horning to avoid the possibility of the trade saying that if Mr. Horning had been there the failure could not have occurred." Other people in the plant voiced similar sentiments, stating that they could not let Horning down, even if he was no longer with them.

Responsibility for the company's failure or success would fall directly on James DeLong. Following his success in the oil field, he had been named plant manager in 1928. His responsibilities had grown over the years, even more so as Horning's health deteriorated in 1935. Promotion from within was the rule of business in the 1930s and Waukesha Motor was no exception. A board member and vice president, DeLong was the logical and unanimous choice to succeed Horning.

Harry Horning, seated at center, at a banquet shortly before his death in January 1936. Elmer Lewis is at his right. James De Long, his successor as president, is standing at Horning's left. The portrait behind them depicts former President Warren Harding.

The Horning Memorial Award

The Society of Automotive Engineers established the Horning Memorial Award in 1938. Its purpose is to recognize research "relating to the mutual adaptation of fuels and internal combustion engines." Over the years, the list of recipients has included some of the most distinguished scholars working in industry and academia on the subjects of greatest interest to Harry Horning, fuels and engines.

Horning presented his first scholarly paper, entitled "Truck and Tractor Engines," to the Society in 1916 and served on the SAE's first Oil and Fuel Committee a few years later. In 2005, the most recent Horning Award was presented to a team of researchers from Korea's Advanced Institute of Science and Technology, Hyundai Motor Company and Sandia National Laboratories in the United States for their work on "The Influence of Charge Dilution and Injection Timing on Low Temperature Diesel Combustion and Emissions."

These latest honorees—an international team of scientists representing research labs on two continents, and a manufacturer—collectively embody "the thinker and doer" that was Harry Horning.

A PIVOTAL YEAR

James DeLong took over the top spot at the start of a pivotal year. With the death of the company's founder, it could not be anything but pivotal. Yet it was also a typical year, in which Waukesha continued to introduce new engines, search for new markets and contend with challenging economic conditions. In 1936, the economy was recovering and, by 1937, the gross national product returned to its 1929 level. Partisans of Roosevelt's New Deal declared that the lyrics of their campaign song rang true—"Happy Days Are Here Again."

Days were certainly happier at the Motor Company. Engines were being shipped at the rate of one thousand a month even though, or perhaps because, the number of models had been reduced to seventy-two—as Horning had recommended in 1935. Employment had more than doubled to sixteen hundred workers and another expansion of the plant was underway. The company newsletter was revived but, instead of the restricted circulation *Turbulence,* it was called *The Waukesha Motor News Bulletin* and distributed in employee pay envelopes. Of greater significance, net income was up to $731,533.12 in fiscal 1936 and on its way to $1,026,825 in 1937.

Orders for motors reflected the recovery. Portable Rig in the oilfields, J. I. Case in farming, Frank G. Hough in construction, Kohler in industry, and a growing number of railroads were buying Model FCs by the hundreds. The FC was selling so well, it seemed that it alone would lift the company out of the Depression. In addition, Oliver was ordering VIMs for farm tractors and Thew Shovel was purchasing several 6B Models for its Lorain-brand excavators. The company also started to make the first of over thirty thousand Model UMA four-cylinder motors for Allis Chalmers.

Model 6LRH

A Model 6LRH powering a cable hoist to move logs over hills and valleys in the Pacific Northwest, 1938. The working conditions were about as rugged as any in the world.

Many other customers made smaller orders. Chrysler ordered several experimental-model diesel six-cylinder truck engines. Standard Oil purchased three 6LRO engines specially modified to run on low-octane gasoline. Standard shipped them to Bahrain in the Persian Gulf to pump crude oil from shore through a twelve-inch underwater pipeline to tankers moored three miles offshore. Stewart Truck, a regular customer since the 1920s, purchased 6BK and XBK Hesselmans for export to the Netherlands. The Washington Iron Works, a

manufacturer of logging equipment, ordered a 300-horsepower 6-LRH Hesselman for a cable hoist that would lift logs to a "skyline" six hundred feet in the air and then fly them two-three thousand feet to a landing.

Waukesha made its first sale in Japan in 1928 and the distributor there, Assano Bussan, continued to handle sales of the CFR and other motors in the western Pacific. In 1932, the Japanese army occupied the northeastern Chinese province of Manchuria. Although nominally independent, Manchuria was now a colony and wide open for Japanese investment. Through Assano Bussan, the South Manchurian Railway had, by 1936, acquired one-hundred-fifty 1928 Model 6RB engines to power individual passenger cars that ran on their own and not as part of a multi-car train. They zipped up and down the one-hundred-fifty miles of track linking Manchuria's principal cities, Mukden and Darien. As reported in the *News Bulletin,* the "motor cars" as they were called, were very popular, and a representative of the railroad was in Waukesha shopping for air-conditioning units for them. If he made the purchases, it is likely that they were the first air-conditioned railcars in China. The Chinese expelled the Japanese from Manchuria in 1945 and the fate of the motor cars—air conditioned or not—is unknown.

When it came to taking Waukeshas to remote places, no purchaser surpassed Ingersoll-Rand. A testimonial to the quality of the engine and of his own equipment as well came from an I-R international sales rep named Frank Ashby. He told the editor of the *News Bulletin* that his company had to be extremely careful in selecting engines for units sold overseas because of the conditions those engines might face. He used the example of a Waukesha/Ingersoll-Rand unit sold to a client in Turkey. Built, tested and hum-

Asian Presence

A Japanese language ad for Hesselmans, 1938. It points out that forty-three million American automotive vehicles were powered by "low-pressure, spark ignition engines," like the Hesselman and therefore the Waukesha Hesselman was a good buy. Waukesha's Tokyo distributor, Asano-Bussan, placed the ad in print media in the western Pacific.

A Waukesha-powered solo commuter rail car running on the Susquehanna Railroad, 1940. Waukesha Model 6RB engines were already powering one-hundred-fifty similar "motor cars" on the South Manchurian Railway in northeastern China.

Rock And Roll

A Waukesha/Ingersoll-Rand compressor unit running rock drills at the site of the new San Francisco Municipal Airport Seabase, 1936. The Seabase would be the home port for the"China Clipper" and other seaplanes that crossed the Pacific prior to World War II. The Seabase was in the United States, but Ingersoll-Rand sold thousands of Waukesha-powered compressors around the world.

ming smoothly in Waukesha, the engine was shipped to the Ingersoll-Rand plant in New York where it was coupled onto a compressor and crated for shipment. Then it sat in the hold of a cargo ship for about a month. Upon arriving at its destination, it was uncrated and dismantled into pieces small enough to be loaded onto pack mules. The mule train set off from the seaport to a site roughly ten thousand feet up in the mountains. There the unit was reassembled and, without a factory rep on hand, expected to run as smoothly as it had back in the United States. It must have worked. Ingersoll-Rand bought fifteen hundred Waukeshas in 1936 and seventeen hundred in 1937.

Ingersoll-Rand was also responsible for Waukesha's debut in the movies. As part of a promotional campaign to market Hesselmans in Africa, I-R wanted to show how the motors were manufactured. Waukesha hired Warren O'Brien, who had made thousands of photographs for the company since the early 1920s, to shoot a moving picture. He obliged and the resulting film was a hit in the home town. It played to a packed house at the largest auditorium in Waukesha, the eight-hundred seat high school gym, on four nights in December, 1936. O'Brien used color film, thereby anticipating the entry of color movies in Hollywood by several years. Once again, Waukesha can claim a first, as the producer of the first industrial film shot in color.

Probably not for the movie, but Waukesha's international reputation brought visitors from around the world to the plant. One was P.I. Androusenko, Chief Designer at the Kharkov Tractor Plant in the Soviet Union. He came in 1936, when the Soviet Union was in the grip of Stalin's brutally murderous industrialization program. Androusenko said he was visiting the leading motor manufacturers in the United States and had already toured General Motors. Although small by comparison with GM, and with his own Kharkov Plant, which had ten thousand workers, Waukesha was on the Russian's itinerary because of its reputation as the manufacturer of the CFR engine. The Soviets had acquired several CFRs and, ignoring the patents, copied the design and built their own. If Androusenko expressed his gratitude, it was not recorded.

Economic stimulus was part of the New Deal, and so was pioneering social and labor legislation. In 1936, Waukesha workers encountered a new government safety regulation when shop workers were ordered to wear safety glasses on the job. They were unpopular, but mandatory. More popular was the contribution the company was obliged to pay to the newly enacted Social Security retirement program that also began in 1936.

The New Deal, along with the depressed economy, had encouraged public opinion in favor of organized labor. Local Lodge 1377 of the International Association of Machinists had been chartered in October 1933, with nearly all of its original two-hundred-forty members at the Motor Company. Emile Jaquet was president when Horning signed the company's first agreement with the local in April 1935. The agreement recognized the Machinists' right to collective bargaining for its members, established a seniority system and set wages and hours.

In 1936, negotiations grew more confrontational. The Machinists wanted a regular work week of forty-five hours with any time above that clocked as overtime and paid at time and a half. All Saturday hours were to be considered as overtime and compensated accordingly. Average pay for shop workers was sixty-three cents per hour ($9.00 in 2006) with some making less, of course, and some making more. The Local also demanded that the company become a closed shop and that union membership be mandatory for all production workers. Membership had increased to about four-hundred and fifty, slightly more than one-third of eligible employees, but the majority was still reluctant to join. Waukesha's management was willing to negotiate on all the union's proposals except the closed shop. The union did not abandon it and the closed shop would remain a sticking point in labor-management relations at Waukesha for many years.

The year 1936 was extraordinary for Waukesha Motor. It had begun with the demise of the man who had founded and steered the company since its birth. What he might have achieved, what his company might have become, had Harry Horning's life reached the Biblical span of three-score and ten, or more, can't be known. Yet 1936 was also a typical year for Waukesha Motor. The founder had died, but the company lived on.

Four-Cylinder Line

Workers on the four-cylinder line, 1938.

THE LITTLE DEPRESSION

The recovery that had enabled the New Dealers to sing "Happy Days" sputtered in 1937 and stalled the following year. The "little Depression" of 1938 was not as damaging as the big dip of the early 1930s but it was severe enough to knock back all the economic indexes. At Waukesha, the bellwether Ingersoll-Rand order fell to 301 motors and net income declined to $451,685, about the same as in 1932. Net income would not cross the million-dollar threshold again until the pre-war defense buildup started to register in fiscal 1941.

The "little Depression" was disheartening because most Americans thought the economy had turned the corner and the recovery would last. The unexpected roller coaster ride down, and the hard memories it revived, strained the relationship between labor and management throughout the country, Waukesha included. The 1936 and 1937 agreements with Local 1377 had resulted in a reduction in the work week to forty-two hours and an increase in the average wage to eighty cents an hour ($11 in 2006). The following year had seen no wage increase and about six hundred workers laid off.

Since wages for most workers were set on a piece rate, they varied according to an individual's productivity. As a worker performed his part of the engine building process, he was timed and his rate of pay set accordingly. The goal for every worker was to "make his rate" and be paid a full day's wage. If he did not, he was paid less.

A worker who could do his job faster than others, and was therefore more productive, was paid more, but only up to a point. Then the company applied the principle of economy of scale. That is, the more units produced, the lower the cost. Therefore the rate of pay per unit was reduced. A worker who consistently exceeded his rate was retimed and his rate adjusted. The piece-rate system, which was used throughout American manufacturing, often led to misunderstanding, rivalry between workers and subterfuge. For example, it was not unusual for a worker to slow down a bit when he was being timed, so his rate would be lowered and easy to make or exceed. The timekeepers were aware of this ploy and adjusted for it, and not necessarily to the satisfaction of the man being timed.

Workers found other ways to game the system. Machinists making a specific engine part were known to run off a few extras on a good day and stash them "in the hay" to make up for an off day when they might not make their rate. As a rule, Waukesha's shop workers were skilled craftsmen who took great pride in their work and their own abilities. They kept track of the procedures they followed to manufacture components and assemble them into engines that ran as designed. The notebooks they used were their own property and contained invaluable information they did not necessarily share with each other or their supervisors. If they were laid off, fired or quit, their notes and knowledge left with them.

Imperfections in the system were easier to tolerate in good times than in bad. In 1939, membership in Local 1377 had risen to about eight hundred of eleven hundred shop workers. Times were still bad and, in their new contract, the Machinists wanted a reinforcement of seniority rights, a re-timing of jobs, a ten-cent-an-hour raise, and a closed shop. The company maintained the position it had held since 1933. It was willing to negotiate on all points except the closed shop. President DeLong said the company believed "it to be the constitutional right of every employee to determine for himself whether he wishes to belong to a union or not." On November 10, 1939, after a long meeting and much discussion, Local 1377 voted to strike. Local President Emile Jaquet told the *Freeman,* "The union intends to stay out on strike until the closed shop demand is granted."

Ever since the United Auto Workers had won the right to organize General Motors by occupying the plant at Flint, Michigan in 1937, unions had perfected the tactic of the shutdown. It was not enough to stay home from work. Union members had to show up for picket duty en masse and block all entry to the plant. Not long after the strike vote, about one-hundred-fifty men lined up in front of the gate on St. Paul Avenue and at least ninety remained at the gate day and night for the duration. Only the watchmen who patrolled the grounds for burglars and fires and the boiler tenders who kept the plant heated were allowed inside.

At the start of the first regular working day after the strike vote, Monday, November 13, virtually the entire membership of the Local lined up along St. Paul Avenue. When the foremen and office staff showed up for work, they were asked not to cross the picket line and no one tried, then. In the afternoon,

St. Paul Stalemate

Local 1377 of the International Association of Machinists went on strike at the Motor Company on November 13, 1939. At the start of the day shift, hundreds of Machinists lined up to block plant entrances while the office staff and spectators waited across St. Paul Avenue.

Solidarity

The Machinists stood close together on the picket line to keep out non-union workers, some of whom petitioned the county court for a restraining order to get back to work. The sign reads "Men will not do this unless they have to."

with only about ninety picketers present, and with assistance from three city police officers, a few office workers attempted to break through the line, but the picketers out-muscled them. The office staff then hired attorneys to file for a court injunction ordering the strikers to allow the white collar people into the plant. The strike ended before the court could rule on the petition.

In the political climate of the late 1930s, neither the federal nor state government was likely to use armed force to break up nonviolent picketing, while local law enforcement did not have the resources to do so. As the law required, the National Labor Relations Board sent an arbitrator who presided over talks between DeLong and Jaquet at the Avalon Hotel downtown. They made little progress at the bargaining table, so both sides took the argument to the press and to the streets. The company took a full page ad stating its case in the *Freeman*. The union held parades and informational meetings. News

1942

Allis Chalmers places first order for 145GZ engines for its M4 high-speed artillery tractor.

1943

Waukesha develops V-12 diesel marine engine and a twelve cylinder "pancake" aircraft engine for the U.S. Navy.

1944

Employment reaches 2,500, including substantial numbers of female shop workers and commuters who do not live in Waukesha.

spread on its own as neighbors talked to neighbors in the small town. Nearly everyone who worked at Waukesha Motor lived in Waukesha and there was not a home in the little city unaffected by events on St. Paul Avenue.

The impasse continued for one month, during which management examined and rejected the idea of breaking the strike by attempting to organize a company union. On December 12 an agreement was reached recognizing Local 1377 as the bargaining agent for all shop employees, even though union membership would not be a condition for employment. Rates and the method for gauging them would be examined and changed when necessary, as would seniority rights and grievance procedures. The standard work week was set at forty hours, with overtime for Saturday, Sunday and holidays. After one year of work, employees would be eligible for one week's vacation. The strike ended in time for all workers to receive paychecks prior to Christmas.

Shutdown

As the days wore on, the picket line thinned, but the plant remained shut down for one month.

Who won? Both sides always said they were willing to negotiate on all points except the closed shop and the agreement specified that they would continue to talk on all those issues. On the closed shop, the union conceded the point. Union membership would not be a condition of employment. At the same time, the company recognized the union as sole bargaining agent for all employees. It was a successful compromise, which meant that each side won and each side lost—a little. In any event, victory or defeat for either side was not the most significant outcome.

Labor activity like the Waukesha strike of 1939 played a role, along with the general prosperity of the post-World War II years, in creating a distinctive feature of American society in the latter half of the twentieth century—the blue collar middle class. Especially in the first three decades after the war, the United States was unique in that skilled and semi-skilled factory workers could expect to own a home and an auto. They could give their children a solid education, have weekends for leisure and paid time off for vacations, and enjoy a secure, comfortable retirement. Nowhere in the developed world was that more likely to occur than the United States, and Waukesha Motor—labor and management—helped make it possible.

1945

World War II ends; Waukesha produced $80 million of motors for the war effort.

1946

Major plant expansion completed to add five acres of manufacturing space.

1947

New family of diesels introduced with patented spherical combustion chambers.

Model 6WAKHM

WAUKESHA GOES TO SEA

IN THE LATE 1930S, WAUKESHA'S HESSELMANN ENGINES BECAME VERY POPULAR IN THE MARINE MARKET, MAINLY FOR WORK BOATS, BUT ALSO FOR PLEASURE CRAFT.

2

1

The Wanderer

1. A Model 6NKHM is hoisted into the engine room of the tugboat *Rival,* 1940.

2. A 6NKHM also powered the 73-foot cruising yacht *Westerley,* out of New York City, 1940. Appropriately, Waukesha marketed the 170-horse-power 6NKHM as the "Wanderer."

3

4

No Overhaul

3. The *Anna K.* was a 58-foot tuna boat operating out of San Diego, California. Her captain installed a 170-horsepower Model 6ELHM in her in December 1936. In October 1939 he told Waukesha's California distributor that the *Anna K.* had been at sea "constantly," including eighteen months fishing in Costa Rican waters. "We have never done any maintenance work to this engine," he wrote. "We seem to get about a year's operation out of a set of spark plugs by cleaning them occasionally...I may have it overhauled sometime in the future."

4. The *Mary* (top) was a Brooklyn-based party fishing boat that took groups of anglers out on Long Island Sound. The *Mary* was powered by a 150-horsepower 6WAKHM, the Hesselman marine version of the 6WAK, one of Waukesha's most versatile pre-war six-cylinder engines.

5

5. Cranes lifting the *Dravo* into the Ohio River, 1940. Powered by a Model 6ELHM, the *Dravo* pushed barges up and down the Ohio and Mississippi Rivers.

Pre-war Plant

The plant in the summer of 1941, a few months prior to American entry into World War II.

WORLD WAR II

When the 1939 union contract expired near the end of 1940, a new set of conditions held sway in Waukesha and the world. Negotiations were brief. Everybody wanted to stop talking and go to work. Although the Japanese had been fighting the Chinese for several years, it was the war that started in September 1939 in Europe that made the whole world sit up and take notice. In the United States it stimulated the economy as the previous conflict had in 1914 but also alarmed Americans much more than had the onset of World War I. American industry was already tooling up to supply the combatants when the fall of France to Germany in the summer of 1940 prompted the Roosevelt administration to begin the largest peacetime defense buildup the United States had seen. The initial price tag exceeded $100 billion. The economic stimulus it supplied brought about the end of the Great Depression.

Waukesha was well positioned to thrive in an economy stimulated by defense spending. In fiscal 1941, before the United States entered the war, net sales grew to $11.4 million and net income to $1.1 million. The industrial sector accounted for the largest volume of sales, followed by oil, agriculture and trucking—all vital to national defense. The shop floor filled with workers again, another plant expansion was in the works, and the shipping department was busier than it had been in over a decade. The parade of new models continued, with some of the most significant motors in company history coming down the line in the years immediately before and during the war.

The 1938 Model ICK was used on the railroad, in the oil field and in industry, often as a starting motor for larger models, like the 6LR "Giant Six" Series. During the war, the Army used it to run auxiliary power units to run electric generators, water purifiers, pumps, flood lights and air conditioners in field hospitals. The Army Air Force used ICKs to power engine oil heaters on aircraft flying out of air bases in Alaska, Canada, Iceland and other cold-weather stations. Rated at 12 horsepower tops—less than most modern garden tractors—the ICK was also installed in what must have been the smallest military Jeep made, a tiny vehicle that weighed in at eleven hundred pounds. Like the infantry grunt who always did his job minus fanfare and fame, the ICK proved its worth many times over during the war.

The 1939 6WAK was probably Waukesha's most versatile engine, even though only about five thousand were built. The 6WAK was modified to burn gasoline, natural gas, LP gas, butane, or oil, and to run with a Hesselman combustion system. A 6WAK could be found powering a heavy-duty Dart truck hauling iron ore in Minnesota, running a work boat for Sunoco Oil in the Gulf of Mexico, or operating a "motor car" for the Susquehanna commuter railroad in Pennsylvania. It also replaced the 6BK as the most popular motor on oil rigs and gas pipelines. The 6WAK continued in production until 1963 when it was modified slightly and renamed the Model F1197. In this configuration the versatile six remained in production until 1997. Of all Waukesha engines, only the CFR has remained in production longer than the 6WAK.

As the W designation indicated, by the late 1930s, Waukesha was running out of letters to identify new engine series. The letters X, Y, and Z and been taken in the 1920s and, while the Y and Z series were out of production, the XAH was still very much alive. The only letters left in the alphabet bin were the U and, maybe, the Q. The Q had been used to identify one and only one experimental engine back in 1913. It could have been revived years later, but that would have broken the unwritten rule of engine identification that forbade duplication. The U could not be used as the first letter of a series because it had long been used to designate a unit, i.e., a motor plus options. Therefore it was the most commonly used second letter, as in Model AU, KU, PU, TU and many others in between. A Model UU just wouldn't work. When Waukesha started building six-cylinder engines in 1924, the alphabet bin was replenished by the addition of the numeral 6, as in 6A, 6B and on to 6Z. Use of the Q was again permitted and nearly one-hundred-fifty Models 6Q, 6QB and 6QL were built between 1925 and 1928. The U was passed over again, for obvious reasons, but so was the V, for reasons unknown. With only twenty-four letters in the bin, it was only a matter of time before the supply ran out.

Pearl Harbor

Waukesha-powered pumps helped to raise the *U.S.S. Oklahoma* after the Japanese attack on Pearl Harbor. With her side ripped open by torpedoes, the *Oklahoma* rolled over at her mooring with a loss of over four hundred lives. A massive salvage effort began in mid-1942, with the battleship not fully afloat again until November 1943.

Model 150

Built for the Crosley, the Model 150 was, like many Waukeshas, modified for military duty during the war.

The 1939 Crosley auto, the car for every man, was powered by the two-cylinder Waukesha Model 150.

The all-numeral Model 4-95 came out in 1931 and, by the end of the decade, numerals and letters in myriad combinations were stamped on the nameplates of every new Waukesha made. The 4 series, named for the number of cylinders followed by the horsepower rating, was out of production by 1943, replaced by the 130 and 160 series. Nearly the entire run of the single model 160 series—twenty-nine thousand, four-cylinder, 30-horsepower motors built between 1940 and 1948—was purchased by Oliver for tractors.

Before the 160 came the 150 series. Waukesha's only two-cylinder, air-cooled engine, the 150 was initially manufactured for the Crosley Motor Company. An Ohio entrepreneur, Powell Crosley had prospered as a marketer of low cost radios and refrigerators in the 1930s and he thought that he could replicate his success in the automotive field. He designed a compact car smaller than the ICK-Jeep. It measured ten feet overall, with an eighty-inch wheelbase and weighed nine-hundred-some pounds. The 1939 inaugural model cost $325-$350, in part because it was sold through Crosley's hardware and appliance store network instead of auto dealers. Waukesha designed and built the 8-horsepower Model 150 and the slightly larger 151 for the five thousand Crosleys built before the government ordered a halt to all domestic auto production in early 1942. It was a measure of Waukesha's flexibility, or a legacy of the Great Depression, that the company was willing to produce an engine for a vehicle that retailed for $325 ($4,500 in 2006).

The Crosley wasn't much of a car, but the Waukesha 150 and 151 each proved to be quite good motors. The Army purchased several thousand of each and used them as it did the ICK but in applications where less power was required.

The Model 145 came out in 1940. It was aimed at the fire truck market, which meant it had to power a vehicle carrying a reasonably heavy load at a reasonably fast speed. In anticipation of the Allied invasion of Europe, the

Army contracted with the Allis-Chalmers Company to manufacture an artillery tractor that could move big guns fast. After the nightmare of World War I's trench warfare, Allied tacticians planned the post-invasion battle in Western Europe to be a war of movement relying on tanks, trucks and motorized artillery. They were looking for a vehicle that could pull a reasonably heavy load at a reasonably fast speed. Allis met their demands by designing the M4 "high speed" artillery tractor to reach speeds of thirty miles per hour unloaded, although a bit slower with a 155 mm "Long Tom" artillery piece on the tow bar.

To power the M4, Allis purchased the largest Model 145, the 250-horsepower 145GZ. After the Allied troops made their famous breakout from the hedgerows delaying them in Normandy in the first weeks after the invasion in 1944, the tank army commanded by General George Patton led the way to Paris. They justifiably received the glory and the fame accorded the victors. Right behind them, and sometimes ahead, were the field pieces pulled by M4s with Waukesha 145GZs under their hoods.

The 145GZ continued its distinguished service in the civilian sector after the war, powering fire trucks and other vehicles and pumping natural gas through pipelines. In 1962, it was modified and renamed the F817. It remained in production until 1997, to complete a lifespan only one year shorter than that of the 6WAK.

The 145GZ typified Waukesha's contribution to the war effort. The Motor Company was already building just what the Army, Navy or Air Force required, be it the 6SRKR for Four Wheel Drive trucks and Harnischfeger cranes, the 130-GS for Ingersoll-Rand compressors, or the 6BZ, 6MZR or 140-GK for Thew/Lorain power shovels. The challenge was to build more of the same faster than ever, making only minor changes when required. For example, Waukesha built on its experience with alternative fuels to develop kerosene-burning versions of models already in production. During the war the company introduced over forty models, most notably the 185GA, and two engine series, the 180 and the 190, but they were more popular after than during the war.

The Army and the Navy did commission Waukesha to produce several experimental motors. Working in cooperation with the Air-Cooled Engine Company in Pennsylvania, Waukesha built the 12-ACG-806 aircraft engine.

Navy Test Cells

An unidentified worker operating a reaming machine on the Navy aircraft engine, 1943. The reamer was one of many modern pieces of equipment purchased for the war that did much more service afterwards.

Test cells built for the Navy aircraft engine program that never got beyond the prototype stage, 1943. Although built for the war, the cells remained in use for thirty years.

Women At Work

Female workers in the parts department, 1945. By the end of the war, Waukesha's work force exceeded two thousand, with women working in all departments. In addition to the larger-than-ever number of women workers, World War II also brought a sizable number of commuters to the Motor Company. Prior to the war, just about everyone who worked at Waukesha, lived in Waukesha and many of them walked to and from the plant.

The Navy had high hopes for this unconventional, air-cooled, twelve-cylinder "pancake" design in early 1942 and funded the construction of test cells at the plant that proved to be much more useful after than during the war. Plagued by delays and the difficulties of long-distance communication with Air-Cooled, Waukesha was able to ship no more than fifty 12-ACG-806s by early 1944, when the program was suspended. The Army Air Force then expressed interest in using the test cells for an experimental auxiliary power unit for its new B-29 high-altitude bombers, but did not follow up before the war ended.

Not long after the defense buildup began in 1940, the Navy approached Waukesha to design and build an "all steel welded diesel" marine engine. At least one prototype of this V-twelve-cylinder engine was built and tested at varying speeds for ten days in early 1943, but the model did not go into production. The company also produced the 8-horsepower Model TGU "Multi-Fuel Spark Diesel Engine." It used Hesselman technology to run an air cleaner, ventilating fan and a generator for combat tanks. A prototype was built and tested in a wooden mock-up of the turret of an M3 tank. The Army accepted the model for purchase in March 1945 but the war ended before the TGU could go into production.

While developing new engines to fight the war was exciting, the majority of the work required by the war was not. It was the same old, same old, only more of it at a faster pace and compounded by tension, worry and fear for friends and family away from home and in danger.

Work orders flooded into the sales department. In February 1942, only two months after Pearl Harbor, Waukesha had $6 million in unfilled orders, already more than fifty percent of all sales in 1941. When the build-up peaked in the spring of 1943, the company had orders for fifty-two thousand engines, all designated as high priority. Four Wheel Drive needed eight thousand engines for military trucks. Allis-Chalmers needed fifteen hundred for the M4. Waukesha Sales needed one thousand ICKs for oil rigs. Deciding which order to fill first was management's job in consultation with a harried Navy ensign assigned to interpret War Department regulations at the plant. No matter which order was filled first, production had to increase dramatically.

To meet these unprecedented demands, the company made plans to ship motors at the unheard of rate of five thousand a month in July 1943. The pace slackened to only thirty-seven hundred motors a month after a few months, but would remain at that level throughout 1944. By the time the buildup was winding down in the spring of 1945, Waukesha had orders for $21 million ($440 million in 2006) worth of engines, $4 million for parts and replacements and $3 million for refrigeration units.

To meet all these orders, purchasing personnel had to locate materials high in demand and short in supply while competing with other manufacturers that also needed them—today. They had to master War Department rules on rationing, priorities and pricing, then arrange for delivery via the overbooked rail system or, if the supplier was nearby, by a trucker running on rationed tires and fuel.

Finding workers was just as difficult. The pre-war buildup had pushed employment to 1,575 by February 1941. Fearing that the bubble might burst as in 1938, the personnel department continued to "add men cautiously" until December, when males eligible for military service left in droves just as more workers than ever were needed. By October 1942, with demand for employees crossing the two thousand mark, the company began to introduce women to the previously all-male shop floor.

Women had worked at Waukesha Motor since its earliest days. A chivalrous H.L. Horning had sent James Remington downtown to buy a chair for the newly hired "stenographer," Chrissy Laing, so she wouldn't have to sit on a box while taking dictation at the Blue Front. Women had continued to hold office jobs but, with a few possible exceptions during World War I, did not work in the shop. The machinist's world was exclusively male. As the pace picked up during World War II, Waukesha followed the standard industrial practice of the time, placing women in supplemental jobs such as maintenance, shipping, parts and light assembly while reserving the machinist's and heavy assembly work for men as long as possible. By the end of the war, women could be found holding down nearly

Oil For War

Waukesha oil field units, Lancastershire, England, 1943. With oil a precious strategic resource, the British began drilling at home. By the end of the war the British were using American-supplied equipment at hundreds of rigs.

any job in the shop, but management remained a male preserve. No matter where they worked, women were to be paid fairly. War Department regulations specified "equal payment for women, provided work performed was comparable with men."

Figures are unavailable for the number of women working at Waukesha during the war, but there were enough of them to require major alterations to the locker rooms, lunch room, the "smoking room for men" and "the accommodations," as one report primly states. Rosy the Riveter was the poster icon of the female industrial war worker. No one drew a sketch of Mary the Motormaker, but an artist could have, because she worked at Waukesha.

Managing increases across the board—in production, paperwork and people—was the perpetual problem during the war. In addition to the Navy test cells, the plant was enlarged and a vacant 52,000 square foot building acquired. Located on Elizabeth Street in Waukesha, it was designated Plant No. 2 and was a rare departure from St. Paul Avenue. As the number of employees grew—reaching a record 2,500 by the end of the war—Waukesha also encountered another new phenomenon, the automobile commuter. By the end of 1942 as many as five hundred employees lived outside of the city of Waukesha. Common in the postwar years, driving as much as ten or twenty miles to and from work was rare before the war. It was especially unusual in Waukesha, where nearly everyone walked to work in the morning, went home for lunch and returned for the afternoon. During the war, even with gasoline rationed and new tires all but unobtainable, so many workers commuted and therefore ate lunch at the plant, that the company was forced to add space for them to simply sit down and eat a sandwich out of a brown bag.

And then it all came to an end—the anxiety, extra effort and sacrifice. Peace came to Europe in May 1945 and to the Pacific the following August. For the manufacturers of tanks, artillery shells, military uniforms and other war-specific goods the war was a bubble that burst when the cease-fire agreements were signed. For companies like Waukesha, whose products were marketable in peace and war, the bubble deflated like a balloon losing air but did not go completely flat. The company was as well-positioned for peace as it had been for war. In the brief book published to mark the company's fortieth anniversary in 1946, President James DeLong penned a preface entitled "Facing Forward" and that is where the company set its sights.

Expansion

Navy Ensign Hart, who supervised Waukesha's work for the Navy, construction manager Bob Ruch and James DeLong conferring at the start of the plant expansion, 1942.

War Contribution

In July 1945, the "Advertising Manager" of Waukesha Motor received a letter from Warrant Officer Arthur Pearson, in command of a "crash boat" attached to an aerial reconnaissance squadron in the Sulu Islands, southwest of the Philippines in the western Pacific.

The war would be over in about one month, but Pearson had no way of knowing it. Instead he and his crew performed their duty of searching the Sulu Sea to rescue the crews of aircraft forced down in the treacherous waters. Their boat was equipped with two Waukesha generator units powered by Model FC engines. After five months of hard use, the engines were in need of replacement parts, but Pearson had none onboard and was many miles away from the nearest Air Force supply station.

So he wrote directly to the Motor Company. He figured that he would receive parts faster directly from the Company than through the Air Force supply chain. And time mattered. If the crash boat could not operate, airmen would die. Pearson and his crew even offered to pay for the parts out of their own pockets.

The Motor Company responded with equal generosity. The parts were shipped promptly at no charge. A few months later Pearson sent a gracious and moving note of gratitude.

The story well illustrates the Motor Company's contribution to victory in World War II. The crash boat units were modifications of Waukesha's FC-powered railroad generator sets. Like most Waukesha engines and power units, they were easily adapted to war time applications. They were not exotic or innovative, but they did the job.

The United States was victorious in World War II because of the heroism of men and women like Warrant Officer Pearson. But World War II was also waged on the industrial front and here the United States was also victorious—thanks to the patriotism and sacrifice of men and women at home who gave their soldiers, sailors and aviators the "parts" they needed to win.

An M4 artillery tractor crossing a prefabricated "Bailey" bridge and hauling a heavy field gun in France, 1944. The 14-ton M-4 was powered by a 200-horsepower Model 145GZ fire truck engine. It could hit thirty miles per hour on its own or pull seventeen tons on its towbar. Allis-Chalmers assembled about 5,500 M4s during the war. A larger model, the M6, used two Model 145GZ motors for greater speed and power.

1

Waukesha At War

Waukesha's contribution to the World War II effort began with the defense buildup prior to American entry into the conflict. The company's greatest contribution came not from new designs, but from increased production or modification of existing engines used in vital support, construction and combat.

The Cyclotron

1. A Waukesha Model MZR ran the truck and a Model SRKR powered the crane building the cyclotron at the University of California, 1942. The cyclotron was used for research that led to the development of the atomic bomb.

2. A modified Model FC was under the hood of the small earth mover (below) Clark Equipment built for the Army Air force, 1943. The dozer was designed to fit in the cramped quarters of the C-47 and other transport planes.

3

2

Model 170 V-12

A federal inspector looks over the one and only Model 170 V-12 "all-steel welded" engine Waukesha built for the Navy, 1943.

4

Model ICK

3. The Model ICK (above, left) was a humble workhorse harnessed to run a wide collection of pumps, blowers and generators. The "Air-borne" unit supplied power for Air Force bombers. The trailer unit ran air-conditioners for portable field hospital operating rooms. The Model TGU was an ICK modified to run a ventilation system for military tanks. It was tested in the wooden mock-up of a tank turret depicted here, but did not enter production.

4. Waukesha's Sterling truck (above, right) celebrating victory in the Labor Day parade, 1945. On display were several generator units, a Model ICK "Basic Engine," a Model LRO "Giant Six," and a Model 6SRKR truck engine.

5. Waukeshas lined up for inspection by the Pentagon, 1944. The 450-horsepower Navy airplane engine is at front left, flanked by Model CFR, COT and CUE test engines manufactured for the United States and its allies. Down the line are the Model TGU tank engine and on to the Model 145GZ, Model WAK and Model 6N.

5

Cook Inlet

Alaska's first permanent offshore oil rig began operations in Cook Inlet in 1965, with power from four Waukesha Enginators.

CHAPTER THREE

STATIONARY POWER

THE EXTRACTION AND TRANSPORTATION OF OIL FROM ITS NORTH SLOPE IN THE 1970S PUT ALASKA ON THE MAP OF MAJOR PETROLEUM PRODUCING REGIONS. BUT THE ARCTIC OCEAN WAS NOT THE SITE OF ALASKA'S FIRST OFFSHORE OIL FIELD. In 1965, a decade before the Trans-Alaska pipeline began delivering North Slope crude to the port of Valdez, a consortium of oil companies—Shell, Atlantic Richfield, Standard of California—erected the state's first permanent offshore oil rig in Cook Inlet seventy miles out from Anchorage. Standing twenty stories above the surface in eighty feet of water, Platform I was capable of drilling thirty-two wells thousands of feet into the bottom of the bay. The conditions the rig was expected to endure—hurricane force winds, forty-foot ocean waves, sub-sub-zero temperatures, pack ice six feet thick, plus earthquakes—were extreme even by the rugged standards of the oil-drilling industry.

Platform 1 lived alone in a rough neighborhood so it was designed to be the most self-sufficient offshore oil rig built. Since severe weather could cut it off from the rest of the world for weeks, it was fuelled and provisioned to keep its fifty-person crew working round the clock for forty-five days. Power was an essential part of the equation. All but a fraction of the platform's operating power—from the lights on the helicopter landing pad at the top of the rig, to the drill tip boring through the mud a thousand feet below, and for the workshops, transfer pumps and crew quarters in between—was generated by four Waukesha Enginators®.

The engine component consisted of four Model L5788 DSIU twelve-cylinder turbocharged, intercooled, heavy-duty, medium-speed diesel engines, each rated at 1,356 maximum horsepower. Three of the L5788s powered pairs of

Model L5788

One of the Model L5788 units destined for Cook Inlet ready to ship.

750-kilowatt continuous, direct-current generators dedicated to the work of producing oil. They ran the draw works, rotary table, mud pumps and other gear at the heart of the rig. Although connected in tandem, a single generator could be switched off and both could be throttled up or down as needed. As one drilling foreman reported, "Our ability to switch from one generator to the other hands us perhaps the most flexible power system ever installed offshore. We're able to get what amounts to the work of eight generators from just six." He was able to run most of his equipment at 500 horsepower instead of the maximum 700, which meant long, continuous, low-maintenance operation.

The fourth L5788 powered a single 500-kilowatt generator that met other demands for power on the platform—maintenance shop tools, radio communications, the oven in the galley. Energy was not wasted. Exhaust heat from the engines was channeled to keep drilling mud thawed despite minus-forty degree temperatures. Heat from the engine cooling systems was used to warm the eight thousand gallons of water in the platform's potable and drill water tanks. In this little world perched on legs in the sea, noise and vibration were important considerations, but not a problem. One oilman said the engine noise from the Waukeshas was "a veritable whisper." They also ran so smoothly that no special vibration mounts were necessary. Configured to burn diesel fuel when installed, the L5788s were also designed for conversion to natural gas as soon as Cook Inlet could provide an uninterrupted flow. The most self-sufficient oil rig in the world became even more self-sufficient.

The L5788s on Platform 1 illustrated the progress Waukesha Motor had made in the two decades after World War II. That it takes a full sentence, instead of a few adjectives, to describe the engine, is testimony to its sophistication and that of the company. The L5788 was a pioneering example of the high-powered, high-tech, stationary power plants that would become Waukesha's bread-and-butter engines in the years to come. The L5788 began life as the Model VLRO in 1954 and was Waukesha's first V-12 production engine. Redesigned and retitled in 1963, the VLRO/L5788 survived for only a few years and for only two hundred engines. They were few in number but significant in impact. In 1966, the L5788 was redesigned again and renamed the L5790. It was the first model of the VHP®, Very High Power, line of gas-fueled engines that would carry the company for the rest of the century.

Twenty years before those four L5788s started to "whisper" above the gray waters of Cook Inlet, the Motor Company was focusing on motive power, or at least on motors for equipment that moved—farm tractors, fire trucks, semi-tractors, towboats, highway trucks, construction cranes, fishing trawlers, front-

WAUKESHA CHRONOLOGY

1948
Postwar recession ends, sales return to World War II level of $26 million.

1949
The term "Enginator" first applied to an engine-generator unit.

1950
Clark Equipment makes its first of many orders for engines to power forklifts.

end loaders, and the rail cars where the first "enginators" were installed. Waukesha was justifiably proud of its versatile new diesel models—the 148DK and 6WAKD—which were reconfigured from older gasoline-burning models—and the newer 180DL and 190DL, which featured a new, patented spherical combustion chamber. The company made a full commitment to diesel engines in the 1940s. In the 1950s, they would play as prominent a role as gasoline motors had in the 1920s.

It was fitting then, that the honor of bearing Serial Number 1,000,000 went to a diesel. It was a Model 135DK shipped to the Electric Service Engineering Company in 1956. The records are not exact, especially from the early years, so it cannot be established that the Electric Service 135DK was the one-millionth engine Waukesha manufactured. However, the records are complete enough to state that sometime in 1956, Waukesha made its one-millionth motor. It was a happy coincidence that the company crossed that threshold in its fiftieth year.

While diesels represented the wave of the future, a limited number of alternative-fueled models still came down the line. Holdovers from World War II, when engines that could run on alternative fuels were in demand, Models 180KL, 185KL and 190KL were Waukesha's last kerosene burners. Nearly all of the roughly 5,000 four and six cylinder KLs produced between 1944 and 1952 were sold to Oliver Tractor. Still interested in kerosene burners, Oliver commissioned Waukesha to build an experimental Model 195KL in 1954, but did not order it into production.

The Hesselman engine sounded its last hurrah with the demise of the XAKH in 1958, but the line was all but silent much earlier. Of the twenty-nine other Hesselman models, none was manufactured after 1951. The Hesselman died, but not for lack of effort from the Motor Company. The 1938 Model 6NK, 1939 6WAK and 1940 Model 145 were reconfigured as multi-fuel burners aimed at the marine market and over three thousand left the dock. However, a diesel engine could do anything a Hesselman could, provided the right fuel was available. By the late 1940s, diesel fuel could be found in all but the most remote locations worldwide.

Gasoline motors continued to sell, with thousands of the Series 135, 140, 145, 180, 185 and 190 shipped to Oliver and Allis-Chalmers for farm tractors.

Oliver Diesel

An Oliver Model 99 tractor powered by a Waukesha Model 195DLB, 1951.

1951

Diesel family expanded to include a complete line of automotive and industrial engines, plus power units in the 16 to 135 hp range.

1952

Waukesha introduces its first turbocharged diesel, the Model 135DKBS.

1953

The new research center begins operation.

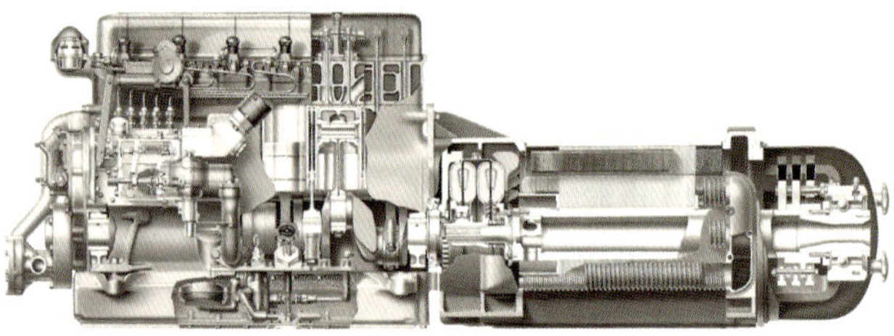

Engine-Generator

The Model 190 DL railroad generator unit was the first Waukesha identified as an Enginator®.

Thew/Lorain and Harnischfeger ordered them for cranes, F.G. Hough for its front end "Payloaders." Ward LaFrance, Peter Piersch and Seagrave mounted them in fire trucks. Four Wheel Drive put them in long haul semi-tractors and heavy-duty work trucks. Kohler was well-known for its generator sets, powered by Waukeshas. Also popular was the 1948 6MZA automotive engine. It found a niche at Four Wheel Drive, Thew/Lorain, Harnischfeger and Waukesha's oil field distributors.

In 1950, Waukesha began a long relationship with Clark Equipment, which had pioneered the forklift as a means to move heavy objects in industrial settings. Clark purchased nearly ten thousand engines in the 1950s, including Model 190 and 195 gasoline engines and Model 135, 190 and 197 diesels.

The railway market looked very promising in the early postwar years. Since 1935 Waukesha had been building "ice engines" for air conditioning and generator sets that provided up to 7.5 kw of power for light, heat and appliances on passenger cars. They were powered by LP-gas-burning ICK and FC motors. After the war, the railroads started trading in their steam locomotives for diesels and were in the market for diesel-powered air conditioners and generators as well. Waukesha promptly coupled a Model 190DL onto a 25 kw generator and called it the Engine-Generator. It didn't take long for the two words to be coupled as closely as the unit they identified and for the marketing department to give birth to the Enginator®. The first unit so named was installed on a Southern Railroad car in 1947 and the first sales brochure published in 1949. When the Milwaukee Road introduced a new line of luxury Super-Dome cars, each one was equipped with two diesel Enginators delivering

To Canada

A Model 6NKDBU is loaded onto a transport plane for duty in northern Canada, 1955.

up to 45 kw. They cooled the car in summer, heated it in winter, and powered appliances year round.

The debut of the Super-Dome was a bright moment in the otherwise dark demise of passenger rail in the 1950s. Waukesha's rail car Enginator turned out to be a fine product with no market. Instead of building new cars, the railroads discontinued service and junked passenger cars as quickly as federal regulations would allow.

L.W. Melcher, who had first brought the ice engine to Waukesha in 1931, was optimistic about adapting the Enginators for refrigerated freight cars. In 1953, he saw a "billion dollar" market for Model 180 and 190 diesel Enginators cooling fruit, vegetables, meat and other perishables on the rails. Since freight cars required a higher ground clearance than passenger cars, company engineers redesigned the older Enginators so the motors reclined at an eight-degree angle above the horizontal. The new models easily fit beneath freight cars, but few of them got the opportunity to do so. Waukesha was not able to penetrate the "reefer" car market dominated by big players like General Motors and Westinghouse. Of the twenty-five thousand railroad units Waukesha shipped between 1935 and the Railway Division's demise in 1956, less than two hundred were diesel Enginators.

Waukesha was able to compete successfully in the automotive "reefer" market. In 1963, the company introduced a redesigned version of the venerable ICK, renamed the ICKN. Eight thousand were shipped to Thermo-King for its highway truck refrigeration units in the following decade. With its market gone, the Railway Division was retitled the Special Products Division and received a new mission as designer and builder of stationary powerplants like the Enginators on Alaska's Platform 1.

Enginator®

Cover of the first Enginator® product bulletin, 1949.

Waukeshas at Work
1950s-1960s

1

2

1. Two Model 6WAKU engines drive generators on a gas compression rig (top) near the mouth of the Mississippi River, 1951. Compressing natural gas offshore, instead of flaring it off, was not yet a common practice and the barge-mounted river rig was an innovative application.

2. Protected by thatched roofs, two Model 135DKBSU Enginators (highlighted, above) sit in their crates in the village of Nauta on the banks of the Rio Maranon in Peru, 1966. Nauta is located three thousand miles up from the mouth of the Amazon. Model 135s in the power house (above, left) a few miles down river at the village of Requena.

3. A 110-ton Dart truck (right) on the Mesabi Range in Minnesota, 1968. The huge truck, with wheels eight feet tall, was powered by an 800-horsepower, Model L1616DSI.

3

4

5

6

4. A Model F2894GU, natural gas fueled engine (above), running a secondary recovery pump in a California oil field, 1965. With its twenty-foot stroke, the pump raised about nine hundred barrels of oil and water a day, from a depth of 7,400 feet.

5. A 35-ton Insley crane (top, left), with a Model 145 under the hood and a Model 135 on the crane, repairs the the ski lift about 8,000 feet up the slope of Squaw Peak at Squaw Valley, California, 1961. The site was used for the Winter Olympic Games in 1960.

6. Four Wheel Drive Company fire trucks (left), powered by Model 145GKB engines, exercise their pumps along the Chicago river, 1954.

NEW PEOPLE, NEW MOTORS

Just as its motors evolved in the postwar years, a generational change occurred at Waukesha Motor. In 1944, Edward Estberg died at eighty-one years of age. The last active member of the 1909 investment group, he was still on the job as company treasurer and board member. Estberg was known as the local banker in Waukesha but, like many people at the Motor Company, his abilities were recognized well beyond the city on the Fox. He served as a director of Waukesha Motor for over thirty years and as a director of the Federal Reserve's Chicago regional bank longer than any other board member of that thirty-year-old institution.

Anton Herzberg, the superintendent who had set the pace for work on the shop floor for forty years, retired in 1962. Starting as a tool and die maker and moving up to shop superintendent, "Tony" Herzberg mastered his trade and, in no-nonsense fashion, conveyed his commitment to efficiency and quality to every worker who came under his tutelage. He insisted, for example, that assembly workers arrange their tools on the tool cart as meticulously as if they were surgeons in an operating room. A tool should be where it was supposed to be and no time was to be lost rummaging through a pile of wrenches and pliers to find just the right one. At least two generations of engine builders learned that there were two ways to build an engine, the wrong way and Tony's way; and Tony's was the right way.

In the engineering department, the clearest sign that the old generation was giving way to the new appeared in 1947 when "Chief Engineer" James B. Fisher retired. He had started with the company when it began its surge as a manufacturer of motors for farm tractors in the opening years of World War I. Not long after Fisher started, the Company developed the Model T, the first Waukesha engine with a detachable cylinder head and a separate cylinder "bucket" cast in pairs. It was one of the first in a long line of innovations that came out of the engineering department on Fisher's watch. He helped develop the Model B military transport motor in

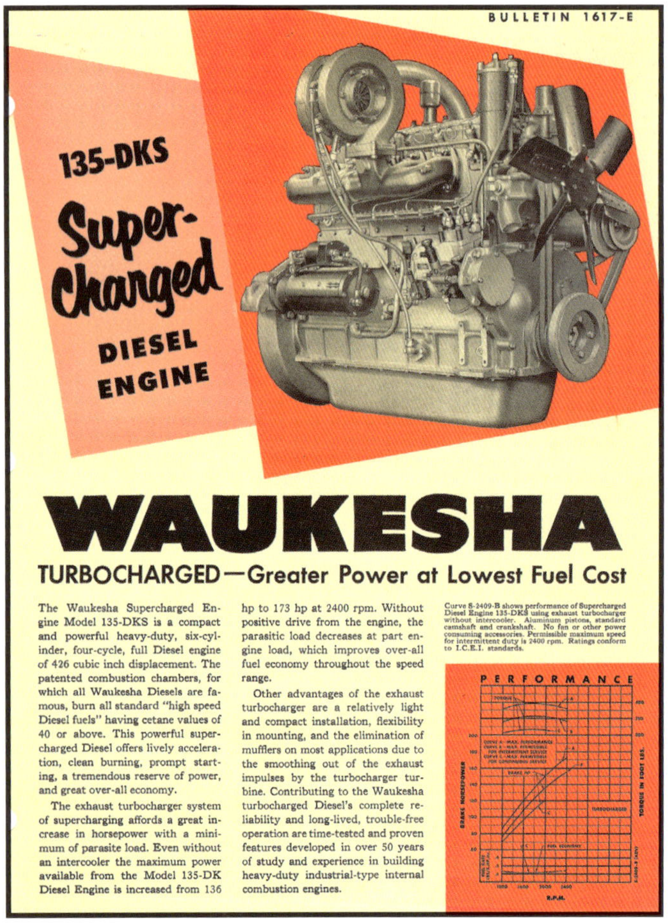

BULLETIN 1617-E

135-DKS Super-Charged DIESEL ENGINE

WAUKESHA

TURBOCHARGED—Greater Power at Lowest Fuel Cost

The Waukesha Supercharged Engine Model 135-DKS is a compact and powerful heavy-duty, six-cylinder, four-cycle, full Diesel engine of 426 cubic inch displacement. The patented combustion chambers, for which all Waukesha Diesels are famous, burn all standard "high speed Diesel fuels" having cetane values of 40 or above. This powerful supercharged Diesel offers lively acceleration, clean burning, prompt starting, a tremendous reserve of power, and great over-all economy.

The exhaust turbocharger system of supercharging affords a great increase in horsepower with a minimum of parasite load. Even without an intercooler the maximum power available from the Model 135-DK Diesel Engine is increased from 136 hp to 173 hp at 2400 rpm. Without positive drive from the engine, the parasitic load decreases at part engine load, which improves over-all fuel economy throughout the speed range.

Other advantages of the exhaust turbocharger are a relatively light and compact installation, flexibility in mounting, and the elimination of mufflers on most applications due to the smoothing out of the exhaust impulses by the turbocharger turbine. Contributing to the Waukesha turbocharged Diesel's complete reliability and long-lived, trouble-free operation are time-tested and proven features developed in over 50 years of study and experience in building heavy-duty industrial-type internal combustion engines.

Curve S-2409-B shows performance of Supercharged Diesel Engine 135-DKS using exhaust turbocharger without intercooler. Aluminum pistons, standard camshaft and crankshaft. No fan or other power consuming accessories. Permissible maximum speed for intermittent duty is 2400 rpm. Ratings conform to I.C.E.I. standards.

Turbocharged

Product bulletin for Waukesha's first turbocharged engine, the 1952 Model 135DKBS.

1954
Waukesha introduces its first production V-12s, the Models VLRO and VLRD.

1955
The "Diesel-Icer" system is Waukesha's final entry in the declining railroad market.

1956
Waukesha celebrates its 50th birthday and issues serial number 1,000,000 to a Model 135DK engine.

1917-18, the Waukesha version of the Ricardo L-type cylinder head in the mid-1920s, the CFR engine in the late 1920s and the F-type cylinder head in the early 1930s. In his final years with the company, Fisher's engineering department introduced a redesigned CFR test engine and the new postwar diesels.

Fisher was gone but his department continued to create engineering landmarks. Among the most significant was the 1952 Model 135DKBS, the first turbocharged, production model Waukesha engine. Supercharging, which increases an engine's horsepower by forcing more air and fuel into the cylinder than natural aspiration, was not new—Waukesha had a supercharged Hesselman engine in 1940—but Waukesha's turbocharging was new. It utilized lost energy in the engine's exhaust gases to drive a turbine-type supercharger to force air under pressure to the intake manifold and increase horsepower. For the 135DK Series, turbocharging meant a twenty-thirty percent increase in horsepower.

A decade later, the engineers went one better by adding intercooling to the turbocharged engine. They determined that they could further increase horsepower by cooling the turbocharged air before it entered the intake manifold by means of an intercooler. The patented Waukesha intercooler was a water-to-air heat exchanger that reduced the temperature of the supercharged air and therefore increased its density. "The over-all result," as a 1963 press release stated, "is greatly increased power output over the capability of a normally aspirated engine, without a proportionally greater increase of the thermal load." Waukesha engines ran cooler and produced more horsepower. On the Model F1905DM six-cylinder marine diesel, for example, turbocharged intercooling boosted output from 240 to 365 horsepower. On the Model L5788DM, it raised output from 725 to 1,110 horsepower. Although developed initially for diesels, the process was adapted for natural gas engines with similar results. Waukesha motors had been pushing boats over the water ever since the Model A chugged across Pewaukee Lake in 1906. Diesels and Hesselmans were marketed for marine applications in the late 1930s and during World War II, but the marine sector never accounted for more than a small percentage of engines shipped. Activity in this realm picked up as the company invested

VC Power

A Pierce-Pacific tow tractor, powered by a Waukesha Model H1077 LP-gas-fueled engine, moves a giant Boeing 747 airliner, 1968. The H1077 was one of the first in the VC family of engines introduced in 1966.

1957
Waukesha acquires the Climax Engine Company of Clinton, Iowa.

1958
Westinghouse licenses Waukesha to build Le Roi engines under the ROlline brand.

1959
Engomatic electronic controls are first installed on Waukesha engines.

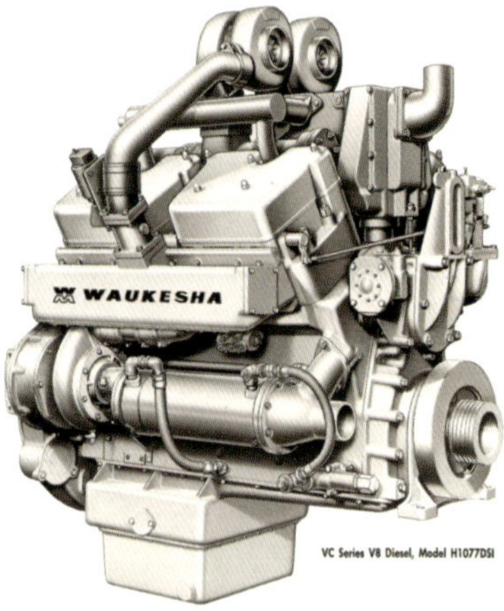

Model H1077DSI

First in the VC Series of engines in 1966, the H1077DSI was also an early example of the new engine identification system: H for eight cylinders, 1077 for cubic inches of displacement, D for diesel, S for turbocharged and I for intercooled.

more resources in larger motors in the late 1950s. Marine versions of the 135DK, 148DK, 6NKDM, 6WAK, and the 6LRD Series were introduced in 1957 and 1958, followed by larger the VLRD and 197DL models in 1959. They were designed for continuous heavy-duty use as primary and Enginator power for commercial fishing vessels on the high seas and for ferries and work boats in rivers and coastal areas.

They led directly to the introduction of the VC Series of high-speed, high-output engines that started with the H1077 V-8 and the L1616 V-12 Series in 1966. Christened the VC or "very compact" line, these engines left a considerably smaller footprint than previous designs without sacrificing horsepower, which is why they were aimed at the marine market where space in the engine room was at a premium. The VC line was also adapted for highway trucks, construction equipment, oil drilling and pumping, and for Enginators.

In 1962, the engine designation system was altered, based on a new set of code letter prefixes and suffixes plus numerals listing the engine's cubic displacement. For example, one of the first engines to be identified in the new system was the 1962 Model F1905GSI. It had six cylinders as indicated by the prefix F, the sixth letter of the alphabet, and a displacement of 1,905 cubic inches. As the G indicated, it was gas fueled. The S stood for turbocharged and the I for intercooled. In 1963, the first Model L5788 appeared. It had twelve cylinders, as indicated by the L, the twelfth letter of the alphabet, and displaced 5,788 cubic inches.

1950s Plant

Aerial view of the St. Paul Avenue plant in the 1950s. In 1946, the Motor Company rented land on "Motor Hill" to the city's Veterans Housing Commission which erected war surplus apartment units to ease the housing shortage for veterans of World War II. The units were occupied until 1960 when they were sold at auction and removed.

To a layman, model designations look as if they were concocted by a church bingo caller pulling coded balls out of a rotating drum, but the system is actually no less orderly than that of any other special language. Once you understand the code, everything is clear—except for the exceptions. Models bearing the suffix R, as in the 1963 L5788GR, might have a cast iron case or might be a redesigned version of an earlier model. Likewise, the S suffix can stand for small bore or turbocharging unless there is a T suffix, which can also indicate turbocharging. Like any living language, *Waukesha-ese* changes with the times.

Customers didn't need to understand the model designation code in order to know what kind of engine they were getting. In the course of purchasing a Waukesha, their needs were thoroughly discussed and they learned which engine would best suit their application and budget. If the motor met these requirements, the model designation didn't matter. Inside the company, the designation code was part of the native tongue. Knowing the difference between an F2894DSIM and an L7040GS was as basic as knowing the difference between a crankshaft and a crankcase. And there were always new models to learn. In 1963, for example, Waukesha introduced twenty-eight new models, followed by twenty-six in 1964, before dropping to only seventeen in 1965.

Engomatic®

Fact sheet for Engomatic® controls, 1971. Predecessor of the sophisticated computerized controls to come, the first Waukesha Engomatic Power Control System was introduced in 1959.

Electronics began to play a role in the predominantly mechanical world of the engine builder in the late 1950s. Many Waukesha engines and generator sets were installed in unmanned locations, such as offshore oil rigs, or in applications where the equipment had to be ready to kick in 24/7 but it was not necessary and certainly not economical to have on operator always present—as in a hospital auxiliary power plant. To meet this need, Waukesha engineers designed the Engomatic® Power Control System. Introduced in 1959, the system could be operated by radio signal, time clock, thermostat, or liquid level as in a gas, oil or water pump. Although the term was not used in 1959, the system could be programmed to make repeated attempts to start an engine that did not turn over on the first try. The timing and number of additional attempts could be set so as not to damage any of the unit's components should it fail to start at all.

The Engomatic system could be called computerized control without the computer, which had yet to be developed. The Engomatic was, of course, the primitive predecessor of the sophisticated computers that became integral parts of every Waukesha engine in the years to come.

Just as the Engomatic control system was pointing the way to Waukesha's future, President James DeLong was bringing a previous era to a close. He

Charles E. Nelson

Charles Nelson, Jr., who became president of the Motor Company upon the retirement of James De Long in 1960.

retired as president in 1960, having held that position for twenty-five years, longer than anyone else. Only Harry Horning had run the company longer, but as president for only about ten years. DeLong had been in management during the 1920s boom, the Great Depression, World War II, and the postwar period of growth. He was a solid business man, seemingly as conservative as his suits, but as aware as anyone that if Waukesha was to do well, it had to adapt. The record of his stewardship is one of adaptation, from Ricardos in the oil patch to turbocharged Enginators in city high-rises.

The company DeLong left was decidedly different from that of the 1920s or the 1940s. The railroad sector was dead, and the automotive, farm tractor and construction equipment markets were declining. Fire truck builders were still buying Waukeshas, but many of the old companies were no longer in business. Waukesha remained in the firefighting sector by diversifying into fire-underwriters-listed power units for driving centrifugal fire pumps in commercial and industrial applications. They didn't have sirens, but they put out fires.

The company's presence in the oil industry was also changing. with natural gas compression units becoming as important as oil drilling and pumping installations. The oil field market was growing and would soon account for nearly forty percent of all Waukesha's sales.

Sales to governments at all levels had increased to their highest peacetime level. Local governments purchased engines for water and sewage pumping and Enginators for auxiliary and special needs. In Canada's prairie provinces, for example, Waukesha seemed to have a monopoly on keeping the ice hard at municipal curling rinks by means of generator sets sold to city park systems.

On the national scale, the Defense Department purchased hundreds of Enginators in sizes ranging from 40kw to 500kw as auxiliary power units on Army, Navy and Air Force bases all across the country. The 150kw Model NKDBS Enginator became the standard power source at Nike missile sites stationed in or near every large northern American city. The Nike's primary mission was to shoot down Soviet aircraft bearing nuclear bombs over the North Pole to the United States, but they were also deployed overseas, along with Waukesha Enginators. For example, four Model NKDBS motors were running 175kw generators at the Nike In Flight Control Center at Taipei, Taiwan. Ironically, diesel Enginators were also sold to the Atomic Energy Commission for a research facility in California. NASA also purchased a unit for its Manned Spacecraft Center near Houston, Texas. Had the power grid blinked during one of the Apollo space flights in the late 1960s, the Waukesha would have switched on.

Hospitals had been buying Waukesha-powered generator sets from other vendors like Kohler since the 1920s but the company did not develop this market into a niche of its own until the late 1950s. The company designed and built backup power systems that could take over an entire hospital in the event of a blackout. Heating and ventilating, air compression and vacuum systems, operating and emergency room lighting, elevators and laboratory refrigerators, as well as communication systems inside and outside the facility, could all be switched over to the Waukeshas. A power system devised for a hospital could be easily adapted to a high school or university campus and then to a skyscraper so emergency lighting and elevators would not fail if the grid went down. Two Waukesha Enginators were installed in Chicago's John Hancock Center, the world's tallest building when it was topped off in 1969.

DeLong's successor as president was Charles E. Nelson, Jr., whose father had served on the board in 1909 and had chosen the site for the new plant on West St. Paul Avenue. Nelson, Jr., started work at the company about a year after graduating from the University of Wisconsin in 1927. In 1933, he was sent to Oakland, California to manage the troubled finances of the Fageol Trucking Company. After a year he returned to Waukesha and rose up the ladder to vice president of production and board member. He served as treasurer and, in 1959, was named executive vice president. Just as DeLong was the obvious choice to succeed Harry Horning, Nelson was the first choice to follow James DeLong.

Piston Line

The piston machining line in the mid-1950s.

HIGH-RISE

WAUKESHA ENGINES HAD BEEN POWERING ELECTRICAL GENERATORS FOR EMERGENCY, BACKUP, OR PEAK DEMAND APPLICATIONS FOR DECADES PRIOR TO THE 1960S, WHEN THEY WERE CALLED UPON TO MEET SOME VERY SPECIAL NEEDS.

1. The 31-story Laclede building in downtown St. Louis started functioning as its own utility in 1970, with two Waukesha Model VHP® L7042G and four Model L5108GSI engines providing all the electrical power the Laclede uses. The building has never been connected to the local utility.

2. The Manned Space Spacecraft Center, Houston, 1964. Three Model L5788D engines were part of the backup power system that insured unbroken communication between Earth and American astronauts in space, including the lunar landing in 1969.

3. Astronaut and Milwaukee native James Lovell and Waukesha's Charles Nelson. The heroism of Lovell and his crewmates on the Apollo 13 lunar mission is legendary, along with his message home, "Houston, we have a problem." Waukesha engines provided backup power at the command center in Houston for all the Apollo missions.

1

3

2

POWER

6

4

4, 5. The world's tallest building when it was topped out in 1969, Chicago's John Hancock Building (above) had two Model VHP5900DSI Enginators for backup power. The Chicago Hancock's sister tower in Boston was similarly protected from power outages, as was Boston's Logan International Airport and the home of Boston's arch-rival in baseball, New York's Yankee Stadium (bottom).

6. The World Trade Center (left) under construction in New York City, late 1960s. Six Model L5792DSI Enginators® provided emergency power for the twin towers.

5

Climax Diesel

A Climax engine on display at a cotton industry trade show in Memphis, Tennessee, 1959. Waukesha purchased Climax, with its factory and foundry in Clinton, Iowa, to expand its line of industrial power units in 1957.

ACQUIRING AND ACQUIRED

When Charles Nelson, Jr., moved into the president's office in 1960, the company was in as fine a shape financially as it had ever been. Net sales had risen to $40 million ($265 million in 2006), earnings per share to $4.33 and dividends per share to $2.50. The five acres of factory space added at the end of World War II were fully utilized, as was the new experimental lab completed in 1953. Employment had risen to over eighteen hundred and would approach twenty-two hundred by 1965.

The forty-hour week was not up for negotiation anymore. It was the standard for non-salaried workers. Instead, the size of the annual raise, longer vacations and additional health insurance and pension benefits were on the agenda. Relations between management and labor ran smoothly with no serious threat of a strike until 1966. Pickets appeared at the gates, but work did not stop. Local 1377 agreed to a two-year contract, instead of the customary annual agreement. The contract also included a cost-of-living clause to offset the inflation beginning to plague the economy. The total package amounted to an increase in the cost of labor of nearly twenty cents per hour. Higher labor costs that were not offset by increased productivity had yet to set off the cautionary alarms they would trigger in the 1970s. In the optimistic 1960s, just about everyone—machinists, managers, shareholders—presumed that the good times would get better, the American economy would not stumble and growth continue indefinitely.

In the past, Waukesha had grown by expanding a line of engines already serving a particular market, by bringing its motors into new market sectors, or by regularly inventing or adopting innovative technology. Although it could have previously done so, Waukesha waited until the 1950s to take advantage of its consistently strong financial position to depart from its traditional practices and to grow by acquisition.

The Climax Engine Company was a manufacturer of gasoline engines founded in Clinton, Iowa in 1900. Over the years it had produced motors for various applications, but had eventually narrowed its focus to industrial

1960
James DeLong retires as president and Charles Nelson, Jr., accepts the position.

1961
Waukesha generator sets installed at Nike missile sites across the country.

1962
The antique script Waukesha Motor Company logo is replaced by a modern W/M.

engines and built a strong position in the municipal services market. It had an excellent reputation for building solid motors in the 100- to 600-horsepower range that, for example, pumped drinking water in Sault Ste. Marie, Michigan, cleared storm water drains in South St. Paul, Minnesota and drove 500kw generators in St. Petersburg, Florida. Waukesha had not made an concerted effort in the municipal services field but hoped to enter it by purchasing Climax.

In Clinton

The Climax Motor plant on the Mississippi River in Clinton, Iowa.

The acquisition was completed in November 1957. Waukesha added the Climax factory and its gray iron foundry on the banks of the Mississippi River in Clinton, along with its three-hundred plus employees and a line of gasoline, natural-gas, LP-gas and sewage-gas-fueled six, eight and twelve cylinder engines.

At about the same time, the Motor Company was negotiating with the Westinghouse Air Brake Company to manufacture a portion of its Le Roi line of industrial engines. Westinghouse wanted its Le Roi plant to focus on compressor motors and therefore licensed Waukesha to build its industrial engines and a pair of V-8 transport engines. Waukesha built only six Le Roi models, about two thousand engines total, nearly all of them the eight-cylinder H884 propane model for municipal services and industry. They were marketed as ROIline engines, to distinguish them from the Le Roi motors that Westinghouse was still making.

In 1960, Waukesha purchased the O & M Manufacturing Company, a builder of radiators and heat exchangers in Houston, Texas. In addition to making radiators and heat exchangers for Waukesha engines, O & M operated as a subsidiary to distribute power units and parts in the southwestern states. Small at first, O & M's sales increased yearly and the plant was enlarged by sixty percent in 1966.

Three years later, the company picked up the Cerlist Diesel Engine Company and moved its operations from Burlington, North Carolina to Clinton. Cerlist was an attractive acquisition because of its patented lightweight aluminum two-cycle engine, the only engine it made. Cerlists came in three or four cylinder models, but Waukesha hoped that aluminum engines would find a niche and the line could be expanded to six, eight and twelve-cylinder models.

1963
New model designation introduced based on number of cylinders, displacement and an alphabetical code.

1964
Waukesha offices enter the computer age with the installation of Honeywell data processors.

1965
The "Turbinator" gas turbine generator line is in production.

By 1965, the company was operating with its main administrative, manufacturing and sales departments in Waukesha, as always, but with Climax Manufacturing and Cerlist Engine identified as separate divisions in Clinton. Sales and service were managed from Waukesha but also from branches in New York City, Tulsa, Oklahoma, and the three California cities of Huntington Park, Bakersfield and Ventura. Four wholly-owned subsidiaries also handled sales and service: Waukesha Motor Western, Ltd., for parts in the United States, Canada and Mexico; O & M in Houston; Waukesha S.A. de C.V. in Mexico City, for most of Mexico; Waukesha of Florida, located in Miami, for sales in Florida. In addition, Waukesha had sixty-eight distributors in the United States and throughout the "Free World," as nations not ruled by Communist regimes were identified during the Cold War.

The recession of 1961 briefly took the wind out of Waukesha's sales, with volume falling from $40 to $34 million. Then the company rode the crest of the 1960's boom, with sales soaring past $53 million by 1967. All fiscal indicators pointed up—sales, earnings, dividends, capital and assets. In 1966, its sixtieth anniversary year, Waukesha increased its commitment to high-speed, high-output, six, eight and twelve-cylinder engines. In addition to the VHP and VC lines, Waukesha introduced the large bore Climax Models H2475 and L3711 natural gas engines. The Cerlist aluminum engine had yet to enter production but Waukesha had signed an $11 million contract to supply "non-magnetic" (aluminum) engines for the United States Navy's fleet of minesweepers. Also in this year, ground was broken for the World Trade Center in New York City. The Center's twin skyscrapers would eventually be supplied with backup power by six VHP Model L5792DSI Enginators.

To The Sahara

Four Model L5100 300 kw generator units await shipment to the Algerian desert in the yard at the Waukesha Pacific Equipment Company in California, 1966.

Waukesha was not the only American business interested in growth by acquisition in the 1960s and the Motor Company itself was a prime candidate for purchase. The first suitor was Cooper Industries, a Texas corporation that came knocking on the door in the spring of 1967. With $145 million in sales in 1966, Cooper manufactured large turbine engines, compressors and tools, and had a firm foothold in the oil industry. Waukesha and Cooper were a good fit and no strangers. Cooper had been filling out its line of power units with Waukeshas for years and was interested in the gas turbine engine the Motor Company had introduced in 1962 as an addition to its line. Cooper began quietly purchasing Waukesha stock, and the price, not known to fluctuate widely, jumped twenty points to $56 in July 1967.

Waukesha's stockholders did not resist Cooper's offer to take control of the company. They were slated to receive the equivalent of $60 per share. The merger agreement was signed in August and the stockholders of both companies voted in favor at special meetings in October. In November, the United States Department of Justice intervened, declaring that the combination of the two companies would have a negative effect on competitive trade in the oil services industry. Cooper and Waukesha were both surprised and unhappy. As Waukesha's Nelson said, "We complement each other. We do not compete." And that was the problem in the eyes of the Justice Department. The two companies complemented each other so well, they would stifle their competitors. Such was the regulatory climate in 1967.

ROIline

A 300-horsepower Model L-3460 ROIline engine at an oil company water pumping station in Texas, 1960, shortly after Waukesha purchased part of the Le Roi engine line from Westinghouse.

Robert D'Amour

Robert D'Amour served as president of the company from 1970 to 1977.

Cooper and Waukesha could have challenged the ruling in the courts but decided not to oppose it. Waukesha board member S.K. Lowry said such a challenge would be "a very slow and very long process that might take as long as five years." The Cooper merger failed, but it sent a signal to other corporate shoppers that the Motor Company was a good buy.

With Cooper out of the game, the Bangor Punta Corporation sat down at the table and anted in by buying Waukesha stock from Cooper. Bangor Punta owed its start and part of its name to the Punta Alegre Corporation, formerly one of the largest sugar producers in Cuba. After Fidel Castro seized power, Punta Alegre lost all its property in Cuba, but hung on to a few million in cash in New York banks, and was allowed to claim a $21 million expropriation loss to offset future gains. In 1964, Punta Alegre merged with the Bangor and Aroostock Corporation, a holding company formed out of the Bangor and Aroostock Railroad, a potato hauler in the state of Maine. As a holding company Bangor and Aroostock could diversify and with the Punta Alegre merger it became the Bangor-Punta Corporation, and the definitive 1960s-style "conglomerate." The term first came into common usage then to describe a corporation assembled through the acquisition of widely diversified companies with few, if any, links to each other.

By the time it made its bid for Waukesha, Bangor-Punta owned thirty-seven separate companies organized into seven general groups. It owned Metcalf and Eddy Inc., a civil engineering company in Boston; Barker Manufacturing, which made materials handling conveyors in Tampa; Knitbrook Mills, a knitted fabric maker in New York; Producers Cotton Oil Company in California; Smith & Wesson, the famed firearms manufacturer in Massachusetts; Starcraft, a boat, camper and motor home builder in Indiana, to name just a few. Sales in 1967 surpassed the $160 million mark, with the ten companies in the Leisure Time group and the seven companies in the Fashion Fabrics group contributing the lion's share. Bangor-Punta became interested in Waukesha as a producer of auxiliary power systems after seeing demand for them nose up after the 1965 New York City blackout. It did not hurt that the acquisition of Waukesha Motor would increase the conglomerate's net income by almost one-third.

A Clark Equipment forklift, powered by a Waukesha VRG engine, hoists a load. Phased out in the early 1980s, the VR family included the last four-cylinder and gasoline-fueled Waukeshas.

Bangor-Punta was so diverse that the Justice Department could not object to its acquisition of Waukesha Motor. Waukesha's management accepted Bangor-Punta's offer of $70 in stock per share of Waukesha stock in April 1968 but the game was not yet over. Another player made a bid. Southern Gulf Utilities constructed, owned and operated water treatment and sewage disposal services for municipalities in Florida and four other states. With net sales of $568,000 in 1966, Southern was relying on chutzpah and leverage to acquire a company with sales one hundred times larger than its own. Nonetheless, Southern offered to pay Waukesha's stockholders what it said was the equivalent of $92-$96 per share in stock and cash. After Wall Street analysts determined that the Southern offer was actually worth something between $40 and $60 per share, Waukesha's shareholders lost what little interest they had in it.

On July 30, 1968, Waukesha Motor ceased to be an independent company and became the sole occupant of the new Energy Systems group of the Bangor-Punta Corporation.

BANGOR-PUNTA YEARS

Bangor-Punta's management philosophy left a great deal of autonomy to the companies making up the conglomerate. Leaving well enough alone at Waukesha was a good idea since the Motor Company registered more sales and greater profit than any other Bangor-Punta company and by itself outperformed nearly all the other groups within the conglomerate. Charles Nelson continued as president until 1970 when he took a seat on the Bangor-Punta board of directors. He is remembered as the courtly president who took pride in knowing the name of "every" employee and who toured the plant at Christmastime to shake every worker's hand, then stationed himself at the factory gate when the shifts changed to greet anyone he missed.

continued on page 122

Ski Power

It was work and not play so Captain Robert D'Amour and crew kept their ties on to test drive a runabout equipped with a Pleasurecraft stern drive engine in the early 1970s. Waukesha made a short-lived return to the noncommercial marine market while part of Bangor-Punta.

WAUKESHA MOTOR COMPANY

SINCE 1906

WAUKESHA, WIS.

Waukesha MOTOR REG. U.S. PAT. OFF 1

Waukesha Logos

1. An early design of the Waukesha Motor Company logo and, 2, a slightly altered later version used for engine cover decals.

3. The Waukesha workhorse logo represented "real horsepower" in the 1930s.

4. Bangor-Punta updated the logo to 1960s style, stacking the letters vertically in two designs, and an engine decal featuring the 1960s Waukesha Motor logo.

5. In the 1970s, Dresser Industries changed the name of the company to Waukesha Engine and created a new logo.

6. In 2001, Dresser, Inc., altered the design and changed the official color to "Dresser blue" (bottom).

Waukesha MOTOR REG. U. S. PAT. OFF. 2

3

WAUKESHA MOTOR COMPANY 4

Waukesha WAUKESHA MOTOR COMPANY WAUKESHA, WISCONSIN 53187

Carlstadt, New Jersey 07072 • Tulsa, Oklahoma 74107 • Huntington Park, California 90255

PETROLEUM INDUSTRY ENGINE POWER

4

WAUKESHA MOTOR CO 4

Waukesha DRESSER 5

DRESSER® Waukesha 6

The Motorworks

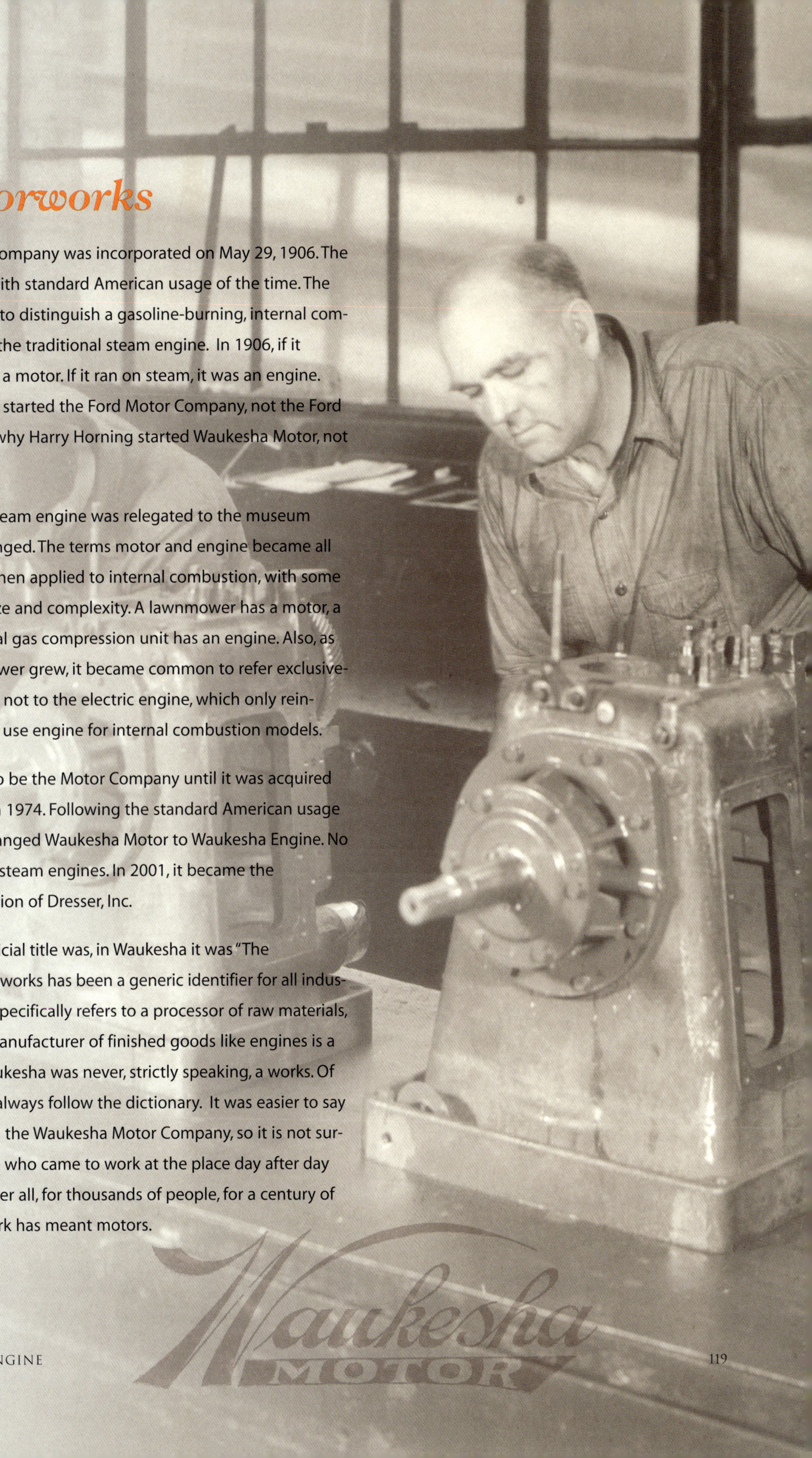

The Waukesha Motor Company was incorporated on May 29, 1906. The name was consistent with standard American usage of the time. The term "motor" was used to distinguish a gasoline-burning, internal combustion machine from the traditional steam engine. In 1906, if it burned gasoline, it was a motor. If it ran on steam, it was an engine. That is why Henry Ford started the Ford Motor Company, not the Ford Engine Company, and why Harry Horning started Waukesha Motor, not Waukesha Engine.

Time passed and the steam engine was relegated to the museum exhibit hall. Usage changed. The terms motor and engine became all but interchangeable when applied to internal combustion, with some distinction made for size and complexity. A lawnmower has a motor, a semi-tractor or a natural gas compression unit has an engine. Also, as the use of electrical power grew, it became common to refer exclusively to the electric motor, not to the electric engine, which only reinforced the tendency to use engine for internal combustion models.

Waukesha continued to be the Motor Company until it was acquired by Dresser Industries in 1974. Following the standard American usage of the time, Dresser changed Waukesha Motor to Waukesha Engine. No one gave a thought to steam engines. In 2001, it became the Waukesha Engine Division of Dresser, Inc.

No matter what the official title was, in Waukesha it was "The Motorworks." The term works has been a generic identifier for all industrial operations, but it specifically refers to a processor of raw materials, as in a steel works. A manufacturer of finished goods like engines is a factory or plant, so Waukesha was never, strictly speaking, a works. Of course, people do not always follow the dictionary. It was easier to say "The Motorworks" than the Waukesha Motor Company, so it is not surprising that the people who came to work at the place day after day used the nickname. After all, for thousands of people, for a century of years, in Waukesha, work has meant motors.

Pink Power

This VHP® Model L7042 required a site-specific paint job to provide back-up power at the Flamingo Hilton Hotel in Nevada in the late 1960s. When it was introduced in 1967, the turbocharged, intercooled Model L7042GSI was the most powerful Waukesha engine built and the top of the line VHP engine.

His successor at Waukesha was not, as might have been expected, a Bangor-Punta transplant, but Motor Company veteran Robert D'Amour. Except for a few years at Cummins Engine right after he earned his engineering degree from Michigan Tech in 1948, D'Amour had spent his entire working life at Waukesha. Starting in the sales department, he moved up the ladder and was vice president of sales when he moved into the president's office.

From his position as chief of sales, D'Amour could see that the company was on the right course and that expansion within the closely linked niches of industrial motors and electrical generator units was the correct heading for the future. In 1971, the Special Products Division was retitled as the Waukesha Power Systems Division, moved to its own facility on Prairie Avenue and given the mission of expanding the company's share of the market for stand alone electrical power systems. The number of sales would be smaller here than, say, in the farm tractor sector, but each engine carried more value and generated more profit. The company could do better by building dozens of the sixteen cylinder, 1,700 horse power 1969 Model P9390G in a year than by cranking

1966
The Model L1616 introduces the VC series of high-speed, high-output, compact engines.

1967
The VHP, Very High Power, series of engines goes into production.

1968
The Bangor-Punta Corporation acquires Waukesha Motor.

out thousands of twenty-horse Model FCs. Accordingly, the Model FC, introduced in 1935, and the 1938 Model ICK, which together had logged over 210,000 units sold, were discontinued in 1973. The swan song for the FC was a short-lived order from Allis Chalmers for a new design to run forklifts in 1970.

New engines were designed or patents acquired to fill all the slots on "four great engine series" and to have "an engine for every application." Design and production were based on the "family" principle, with the maximum number of interchangeable parts to reduce production costs, parts inventories and service problems. In the VHP® and VC Series, for example, two-thirds of all parts were interchangeable within the family.

The VHP Series was at the top of the line in terms of power, with engines in the 280-2100 horsepower range. They were the biggest of the big—six, eight, twelve and sixteen-cylinder powerhouses designed to combine "high specific output with the traditional values of long operating life, good fuel economy and ease of maintenance." They were meant to be the "prime mover" for reciprocating and centrifugal compressors, turbine and centrifugal pumps, oil rigs and electrical power plants. This series included the 1966 V-12 Model L7042G and the turbocharged, intercooled 1967 Model L7042GSI, the most powerful Waukeshas yet built.

Next was the VC Series, in the 420-1,190-horsepower range. These engines were tailored to fit many of the same applications as the VHPs, while taking up less space in the engine room. They could be coupled to generators turning out as much as 660kw and used in commercial and industrial power systems and in marine applications. The 500-horsepower VC Model H1077DSIM powered tugboats on the Gulf Coast while Navy minesweepers plying the waters of Vietnam relied on 700-horsepower VC Model L1616DSINs to turn their propellors and 400-horsepower Model L1616DNs to run machinery on deck.

To meet demands for engines in the 100-360-horsepower range, the company went to Sweden in 1971 to obtain a license to build and sell Scania diesels. Dubbed the VS Series, Waukesha's Scania line originally included four 4-cylinder models, eight 6-cylinder models, and the 8-cylinder Model H866S. Applications were similar to the VHP and VC series, although the VS motors were also marketed to construction equipment manufacturers, mobile working rigs and fishing trawlers.

The least powerful engines were in the VR series. Here could be found examples of Waukesha's diminishing line of gasoline-burning engines, the six-cylinder Models VRG 232, VRG 265, VRG 283 and VRG 310, and one of the company's last four-cylinder motors, the VRG 155. They were aimed at the

Scania H866

A Scania Model H866 eight-cylinder industrial unit sits on a dock, in the 1970s. The setting is appropriate because Scania also had a line of marine diesels. Waukesha purchased licenses to build Scanias in 1971 and marketed them as the VS family.

1969
The VHP Model P9890G is Waukesha's first 16 cylinder gas engine.

1970
Charles Nelson, Jr., retires as president and Richard D'Amour accepts the position.

1971
The Special Products Division, renamed Waukesha Power Systems, moves to an off-site location.

construction, farm, mining and material equipment markets that were becoming increasing less important to the company. In fact, all the VR engines were also configured to burn natural gas for stationary applications.

Model L1616

The 1967 version of the Model L1616 marine engine that was modified for the U.S. Navy.

Through its Leisure Time group Bangor Punta was a sizable manufacturer of powerboats and sailboats with auxiliary motors. Waukesha had built many marine motors over the years, but not since the days of Harry Horning and Fred Ahrens had the company built motors for pleasure boats. Yet, if there was a place where two of Bangor-Punta's diverse companies could meet it was on a motor boat. In 1970, the Clinton plant began to manufacture a modified Ford V-8 engine to run "inboard-outboard," stern-drive systems in runabouts for Bangor-Punta's Pleasure Craft division. Clinton started to manufacture the units, which coupled any of four "Waukesha" branded engines to stern drives manufactured by Volvo-Penta and other manufacturers only to run aground during the economic downtown of 1969.

The American economy had been rolling along like an unstoppable juggernaut since the late 1940s. Recessions had hit in the 1950s and in 1961, to be sure, but they had been brief and all but forgotten in the go-go 1960s. By the end of the decade, the negative effects of inflation were taking their toll, followed by the curious phenomenon of recession and inflation termed "stagflation." The shock of the first oil crisis in 1973 further pushed the economy over the edge. Bangor-Punta was not the only corporation to feel the squeeze, but it was more vulnerable than most. While Bangor-Punta was recording losses, Waukesha Motor had increased sales by twenty percent between 1970 and 1973, when sales reached a record high. Struggling as it was, Bangor-Punta was not able to reinvest Waukesha's profits in the new facilities and technology needed to implement the the Motor Company's plan to focus on large industrial engines and Enginators.

Aware of the situation as early as 1970, the company had acquired two-hundred-forty acres in the village of Hartland, Wisconsin, about ten miles away from Waukesha, and announced plans to construct a new plant and relocate. In December 1973, Bangor-Punta's Chairman, David Wallace, said his company didn't have the money to put up a building for Waukesha. He went on to say that the Motor Company itself was in fine shape and that he would be interested in a joint venture with a partner willing to bankroll the construction. Also, if a buyer came along with the right offer, Bangor-Punta would not say no. Robert D'Amour and several other company executives made an offer of $21 million ($84 million in 2006), but Bangor-Punta's managers were not interested. They had received a better proposal from Dresser Industries and, in early 1974, accepted it.

The USS Avenger detonates a mine, 1990s. Commissioned in 1987, the Avenger was powered by the upgraded version of the 1966 Model L1616.

Minesweepers For The Navy

The minesweeper's mission is to clear safe passageways through fields of magnetic mines that detonate upon the approach of steel-hulled vessels. Therefore, since World War I, minesweepers have been among the few modern naval ships built of wood and other non-magnetic materials.

In the mid-1960s, the United States Navy began to modernize its aging fleet of "ocean minesweeping ships." Mines were now so sensitive to the presence of magnetic iron or steel, that the Navy wanted the seven engines installed in each of its new ships to be built of alternative materials. Six engines would share the work of driving the ship through the water and powering the equipment that detected mines via sonar then detonated them. A seventh, smaller engine would supply electricity for day-to-day operations.

The Navy wanted four to five hundred engines—too few to justify ordering an entirely new design—and instead asked for bids from manufacturers willing to modify a civilian engine for military use. Waukesha had built many engines with aluminum and other non-magnetic components over the years and had a new platform in the VC family, the Model L1616 diesel, that could be modified to meet the Navy's specifications. In November 1965, Waukesha was awarded the contract to supply four-hundred-twenty-nine, 600 horsepower non-magnetic engines to the Navy and began building a modified version of the Model V-12 L1616DSI.

Waukesha shipped three-hundred-twenty Model L1616s before the Navy lost funding for the project in 1973. No more than one-hundred-twenty engines were installed in ships while the balance went into mothballs for the rest of the decade. In the meantime the commercial version of Model L1616 was in trouble. The engine did not live up to expectations and many customers were unhappy with it. In 1975, Waukesha halted production.

In 1981, Congress restored funding for the minesweeper program and the L1616 engines came out of storage. Waukesha corrected the problems plaguing the engine but not before a number of the original models were installed in ships. They did not perform as well as expected. With the Model L1616's reputation tarnished, the Navy accepted a bid from an Italian manufacturer to build engines for fourteen ships. The Italian engines then failed to pass endurance tests that both versions of the L1616 passed. In the end, part of the minesweeper fleet was equipped with Italian engines, part with Waukeshas.

The minesweeper program concluded in 1993, with the last Waukesha-powered ships sold to the Royal Naval Forces of the desert kingdom of Saudi Arabia. Two years later, Waukesha stopped building diesel engines, thereby removing itself from the marine market, nearly ninety years after the first Waukesha motor powered a runabout on Pewaukee Lake.

Bigger Is Better

A 1980's Model 12VAT25GL. Converted from diesel fuel to natural gas and rated at 2,500 horsepower, the 12VAT25GL was an early example of Waukesha's Dresser-era strategy of building increasingly-powerful, exclusively-gas-fueled engines.

CHAPTER FOUR

THE DRESSER YEARS

THE PROCESSING FACILITY AT LONE CAMP, TEXAS, HAS BEEN TAPPING THE NATURAL GAS RESOURCES OF THE FORT WORTH BASIN SINCE THE 1960S. IN 2001, OKLAHOMA-BASED DEVON ENERGY ACQUIRED THE PLANT, RAN IT FOR A FEW YEARS, BEFORE SELLING it to Enbridge. Lone Camp has six Waukesha VHP engines. Four 1,000 horsepower Model VHP L7042GSI engines are coupled with Dresser-Rand RDS-4 compressors to take in gas at 90 pounds per square inch (psi), compress it to 450 pounds psi for processing, then boost it to 600 pounds psi for transmission throughout the American heartland. Engine number five is a VHP L7042G. It powers a chiller that drops the temperature of the gas to remove marketable hydrocarbons. The newest Waukesha engine at Lone Camp is a VHP Series Four L7044GSI. Powering an Ariel JGK compressor and rated at 1,680 horsepower, the L7044GSI moves twelve to thirteen million cubic feet of the plant's capacity of thirty million cubic feet per day. Since it takes about one million cubic feet of gas to meet the needs of about one thousand average American homes, the Waukesha L7044GSI pumps enough gas to power a small city, every day.

In addition to being newer and more powerful than its predecessors, the Series Four engine is equipped with an Engine System Management control unit (ESM). It monitors engine performance around the clock, allowing a technician to check its status via a computer that might be plugged directly into the control unit or located miles away and linked by a wireless connection. Thanks to continuous monitoring and built-in reliability, Lone Camp's L7044GSI logged over five

To Power A City

Devon Energy's Lone Camp, Texas, gas compression station where Waukesha VHP engines process enough gas to power a small American city every day.

Cogeneration

One of two Model16 VATGL power units at the municipal wastewater treatment plant inTampa, Florida. Each unit generates 2.9 MW of power for operations and sale to the local utility. Exhaust heat from the engines is used in operations and further reduces the plant's energy costs.

thousand hours in less than one year and pumped billions of cubic feet of natural gas. With no down time except for preventive maintenance, the engine attained a reliability rate of ninety-eight percent. It was operating at about as close to perfection as an engine could.

At Lone Camp and other gas compression stations, reliability is paramount. The mission is easy to state: keep the gas moving day after day after day after day. Other engine applications are multi-dimensional. In 2001, the Tampa Electric Company completed the installation of two 16V ATGL engines at the Florida's city's wastewater treatment plant. Rated as high as 4,500 horsepower, each of the Model ATGLs power a generator turning out 2.9 MW of electricity. The units are wired to Tampa Electric's grid and help the utility meet peaks in demand, usually in summer. They can also supply emergency power for the treatment plant, should Tampa Electric's service be interrupted, say, by a hurricane.

All engines produce heat. It can be treated as waste and exhausted off or as a useful byproduct. In Tampa, heat from the ATGLs is not just useful, but essential to the application. The 700 degrees of heat the engines produce is the required temperature for drying the sludge created at the wastewater plant. The heat from the ATGLs is vented into the turning-drums that transform the sludge into fertilizer pellets, thereby reducing the plant's use of purchased natural gas. It also reduces emissions because the plant is not burning gas to dry the sludge, which benefits not just Tampa, but the entire planet. The system requires cooperation among all parties, the city, the utility and Waukesha's distributor, Reagan Equipment. The mission is multi-dimensional: wastewater is treated, electricity is generated, two forms of would-be waste—process heat and sludge—are put to use to benefit taxpayers, ratepayers and the environment.

OLD PLANT, NEW PLANT

The Series Four VHP at Lone Camp and the two ATGLs at Tampa are end-of-the-century applications. They are the result of decisions made and strategies pursued in the quarter century since Waukesha Motor became part of Dresser. Although the relationship has changed over time, Waukesha's link to Dresser has been consistently positive for the motor company. When Dresser acquired Waukesha in 1974, neither company was a stranger to the other.

WAUKESHA CHRONOLOGY

1972
Scania engines in the 100-360 hp range complete the Waukesha line of diesels, which ranges from below 100 hp to over 2,000 hp.

1973
Waukesha introduces its first 16-cylinder diesel, the Model P2154DSI.

1974
Dresser Industries buys Waukesha Motor and the Company is renamed Waukesha Engine Division of Dresser Industries.

Founded in 1880 by an inventive and persistent entrepreneur named Solomon Dresser, who developed a leak-proof coupling for oil and natural gas pipelines, Dresser Industries had, by 1973, grown into a billion dollar energy service and industrial equipment manufacturing company.

Divided into five "Segments," Dresser manufactured machinery and equipment for construction, mining and refracting, as well as a diverse line of specialty tools, instruments and controls. Its largest Segment, Petroleum Operations, produced tools, equipment and services for all phases of the petroleum and natural gas industries, from exploration to pipeline, including a line of natural gas compressors that neatly matched Waukesha engines in size and output. Unlike the struggling Bangor-Punta, Dresser was thriving in the inflation-plagued, energy-crisis-ridden economy of the early 1970s. Between 1970 and '75, Dresser's net sales and service revenues had increased from $789 million to over $2 billion ($7.3 billion in 2006 dollars). With its gas compression engines and electrical power systems, Waukesha easily merged into Dresser's Energy Processing and Conversion Segment and helped the Segment earn $449 million in revenue in 1975, its first full year in the company.

Dresser brought a blast of fresh air to St. Paul Avenue. First to change was the name of the company. Following contemporary usage, Waukesha Motor became Waukesha Engine or, more fully and formally, the Waukesha Engine Division of Dresser Industries.

President Robert D'Amour retired from the company in early 1977. He was succeeded by Chet E. Lee. Born and raised in London, England, Lee earned his degree in mechanical engineering and began a career that took him around the world, including an eighteen month stop as chief of manufacturing at Waukesha in 1970-71. He and D'Amour had hoped to modernize the plant or build a new one, but could not find support for their plans at Bangor-Punta. Lee then moved on and, in the mid-1970s, accepted an offer from Chrysler to supervise the construction of a manufacturing plant on the outskirts of Tehran in Iran. His work was finished there when Dresser invited him to return to the United States. He was safely installed in the president's office in Waukesha a few years later when the Islamic revolutionaries led by the Ayatollah Khomeini seized control in Iran.

Chet Lee was the first chief executive at Waukesha who had not spent the bulk of his working life at the company. He started his tenure by calling all twelve hundred employees to meet at the Waukesha County Expo Center, the only convenient space large enough for the seven-hundred-fifty people on the

Chester E. Lee

"Chet" Lee was president of Waukesha Engine from 1977 to 1984

1975
Dresser Industries sales top the $2 billion mark for the first time. The Energy Processing and Conversion Segment, including Waukesha, accounts for $449 million.

1976
Waukesha announces a major plant modernization, including a new high rise warehouse, machinery wing, test cells, tools and equipment.

1977
Chester E. Lee is named executive vice president of Dresser Industries and president of Waukesha Engine.

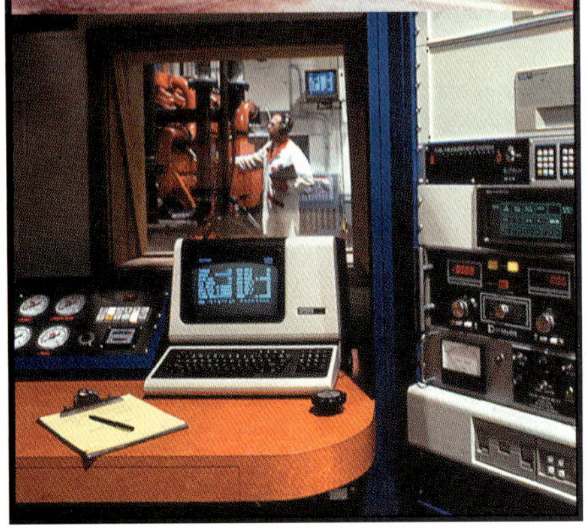

Test Cell

An engine moves into one of the new test cells that were part of the renovation of the plant that took place in the late 1970s and early 1980s.

The new test cell computer station was state of the art in 1984 and one example of the digital equipment that was becoming an integral part of the workplace.

day shift. Actually, Lee held a series of meetings, one for each of the three shifts, with the last starting at 10:00 PM. The new president wanted to introduce himself and state that he and other managers were cautiously optimistic about the future. The first energy crisis of 1973, the Watergate political crisis, defeat in Vietnam, plus the recession of 1973-'75, had been hard on the American psyche and the economy. At Waukesha Engine, more than three-hundred-eighty people were laid off or dismissed in 1976.

Now the signs were more positive. Orders for engines, which had dipped thirty percent in '76, were rising at a good pace. Dresser was optimistic enough to commit $30 million dollars to a multi-year renovation and modernization program that would bring the plant up to current standards. "They were back in the 1920s," Lee recalled. "I'm exaggerating, but still, much needed to be done." With Bangor-Punta limiting capital investment, Waukesha had been slow to acquire the advanced technology and equipment, such as computer numeric controlled machine tools, that were changing manufacturing in the late 1960s and early 1970s. Now was the time to do so.

The renovation plan, as it came to be known, had five phases. It called first for the construction of a "Hi-Rise" warehouse with a computer-controlled parts management system. Under the old system, every part was picked by hand, placed in a basket and loaded on a "trolley" that carried it to the shop. Spare parts for service were in their own department in a building separate from manufacturing parts. "They operated like another company," said Lee. More parts were stored on the shop floor to be on hand for engine assembly. The entire system was labor-intensive and out-of-date. The procedure for keeping track of inventory, for example, hadn't changed since the company was founded. It still required that the company shut down for two weeks while every crankshaft, piston ring and head bolt in the plant was counted, with resulting loss of productivity.

In the new system each part was logged into the computer as soon as it was made in the shop or landed on the receiving dock. It was automatically taken to its assigned spot in the Hi-Rise warehouse. To retrieve it, a worker logged an identification number into a computer and the automated system located and retrieved it. Inventory control was automatic. Shortages were detected promptly. No longer did the plant have to shut down for two weeks every year so each piece could be counted.

The Hi-Rise parts management system was up-to-date for its time but, of course, times changed. By the end of the century it had been replaced by another computerized system. Service parts are currently handled by an independent contractor located off-site. It is a measure of the pace of change in the twentieth century that a manufacturing company could do quite well with the same system for fifty or sixty years, while the newer system that replaced it survived less than twenty.

The renovation plan also called for an entirely new set of test cells. "The test area was in a room off the manufacturing area," said Lee. "There would be machines lined up...when they ran the floor shook and nobody could talk.

Everything was done by sign language. There was obviously a great need to upgrade." Initially, fourteen new test cells were built, with eight more added in the mid-1980s.

Next came new cranes, hoists, shipping and receiving docks to handle the increasingly heavy engines the company was now making. The new lifting machinery was part of a rebuilding of the manufacturing department, including the realignment of the assembly lines. Since they were first laid out, the assembly lines had run north-south, roughly parallel to St. Paul Avenue. In the new plan, the flow of work was shifted to run east-west. The assembly line conveyors, with their rows of engines coming down the line from one work station to another, were replaced by air pallets that raised the heaviest engines off the floor and allowed them to be moved in any direction with a slight touch of the hand. In addition, the assembly areas were sealed off and pressurized to keep dust and dirt out of the engines. "People think it's a dirty business," said Lee, "but cleanliness is absolutely essential for engine building."

Also included in the plan was a modernization of the paint room to meet new health and safety standards. The entire building was retrofitted for energy conservation. Window openings in the walls and in the saw-tooth roofs were sealed or refitted for energy efficiency and the heating/venting/air conditioning system was improved.

In the final phase, completed in 1981, a new front entrance was constructed on St. Paul Avenue and the administrative and engineering offices were remodeled. New technology appeared here as well, with expansion in the use of personal

Assembly Line

Model VR engines destined for Clark Equipment on the line prior to the renovation (bottom).

Engines on the line after the renovation.

computers and the information, accounting and other administrative software that ran them. The new electronic equipment was put to use implementing Edward Deming's "just-in-time" operational philosophy that became standard in American manufacturing in the 1980s. The following year, Waukesha's engineers and drafters started trading in their once state-of-the-art, arm-type drafting machines and mechanized drawing boards for the earliest versions of the soon-to-be wondrous capabilities of Computer Assisted Design (CAD). When the renovation was completed, the plant was an up-to-date facility. It was less labor intensive and idiosyncratic; more automated, computerized and productive.

For Lee and other managers in Waukesha, the renovation was actually Plan B. Had they been able to convince Dresser to support it, they would have moved the manufacturing of the ever-larger engines coming on line to an entirely new plant built in Hartland. In time, all manufacturing would relocate to the new facility. Having just purchased Waukesha, Dresser's managers chose to renovate and keep Waukesha Engine in Waukesha. In time, the engine market and new technology combined to validate their decision. As Waukesha made fewer models and fewer engines, and as it improved management of its supply lines and inventory, less plant space was required. No longer did Waukesha need an entire shop devoted to the manufacture of thousands of Model FC motors, for example. The 9.125-inch-bore, 900-horsepower, 1976 Model F3335, the 1,600-horsepower, 1976 Model L6670, and the 2,000-horsepower, 1982 Model P8894 VHP diesels marked Waukesha's future course, i.e. towards fewer, more powerful, more-value-added engines.

Automation and computerization meant fewer people could be more productive in less space. In the past, Waukesha had increased productivity by expanding its plant until it covered one million square feet. By the end of the 1970's-80's renovation, Waukesha had increased productivity not with more bricks and mortar but by remodeling its plant to utilize up-to-date management practices and industrial technology.

Perfect Image

An engineering department drawing of a Model 6EKH from the 1930s (opposite) reflects the painstaking care taken to draft perfect representations of engines. The drawing is two-dimensional but it still clearly depicts the interior and exterior of the engine. Introduced in the 1980s, computer assisted design technology changed the entire drafting and design process.

1978
The VR Series goes into production.

1979
As part of the modernization program, the assembly lines are realigned from their 1909-north-south axis to a more efficient east-west line.

1980
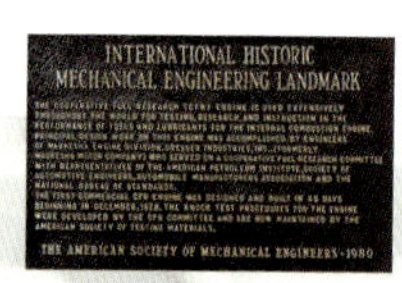

The American Society of Mechanical Engineers designates the 1929 Waukesha Motor Co. CFR engine an international historical landmark.

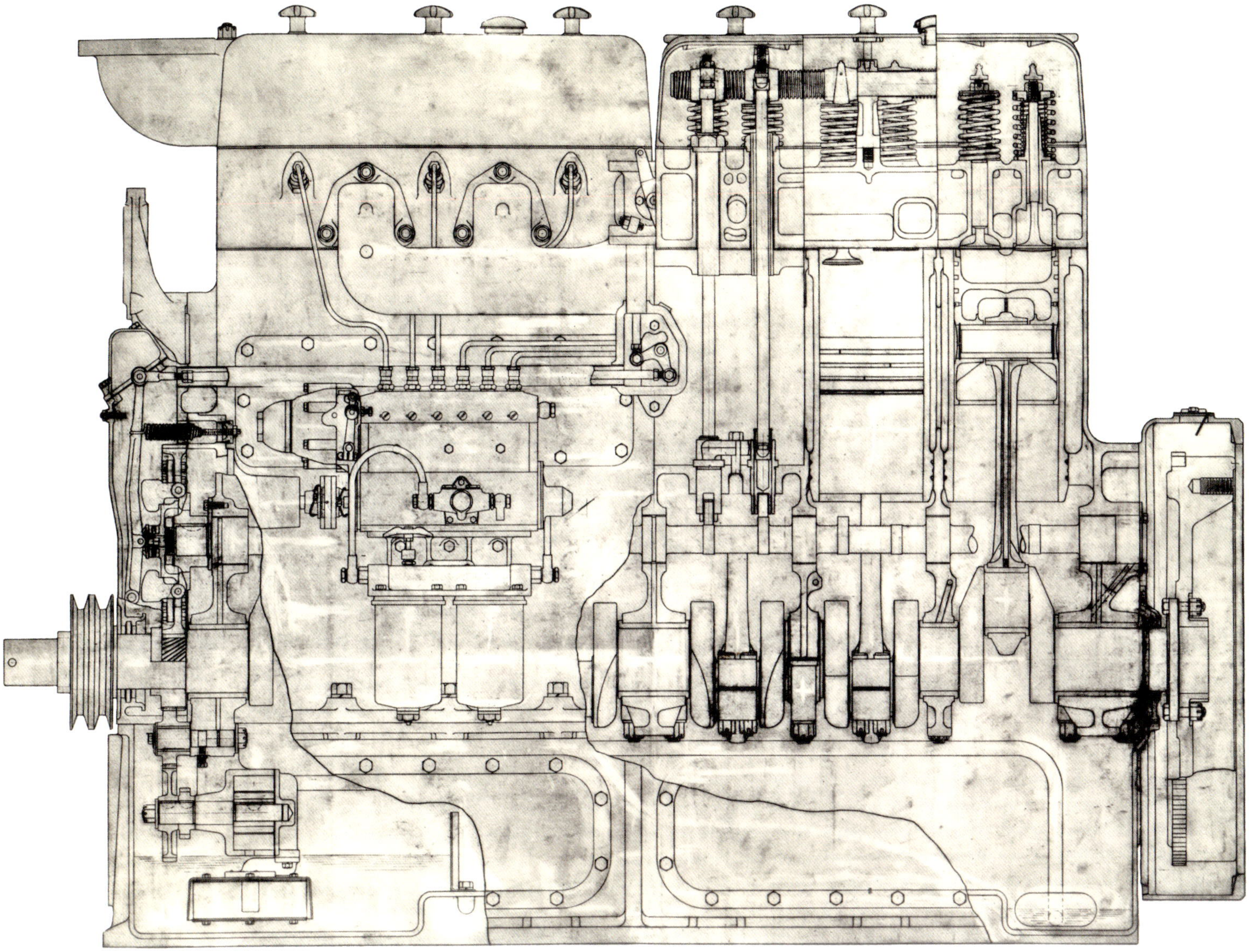

1981
Waukesha purchases licenses from Sulzer Brothers, Ltd., a Swiss manufacturer of high horsepower engines to produce the AT family of engines.

1982
Layoffs begin that reduce the blue and white collar work force from 1,352 to 550 by mid-1983.

1983
Ben R. Stuart is named president of Waukesha Engine.

1

2

Distributors

1. Reagan Equipment offices, Harvey, Louisiana, 1960s.

2. Model VHP 5904LTD Enginators at the Blessed Textile Mill in Lahore, Pakistan, installed and serviced by Masco Energy Services. Masco in Pakistan and Dana Engineers International Ltd. in Bangladesh have helped bring Waukesha Engine to prominence on the Indian subcontinent. Since the early 1980s the two firms have supplied thousands of Enginators that are powering textile mills and a multitude of other industrial applications where continuous power is essential.

3. Service trucks at Pamco's headquarters, Calgary, Alberta, Canada. As Waukesha's only distributor in Canada, Pamco's reputation for superior service helped spur sales of Waukesha engines in the Canadian oil fields and made Canada one of Waukesha's largest markets for gas compression engines.

In its early years, the Motor Company made little use of distributors to reach customers. Not until 1924, when James DeLong met Louis Pearce, Sr., in Houston did Waukesha expand its marketing to take advantage of the services that people who knew their territory could provide. Pearce had founded the company he called Portable Rotary Rig in 1923 to supply service and equipment to the oil industry, including gasoline engines to run drilling rigs and pipelines. Pearce and Waukesha meshed so well and developed such a high regard for each other that, in 1933, Portable Rotary Rig became Waukesha's exclusive Texas distributor. During World War II Pearce, changed the name of his company to Waukesha Sales and Service. As the company expanded into the construction equipment, power package and energy service industries, it was reorganized as Waukesha-Pearce Industries, retaining the name of and maintaining the relationship with its oldest partner.

The tie between manufacturer and distributor is one of mutual dependence. The manufacturer relies on the distributor to make sales and to provide maintenance, parts and repair services to all the company's served markets. They support the "power partners" who design and construct gas compression facilities, such as the Aktobe Gas Treatment Plant in Kazakhstan, which operates six Model 12V-ATGL engines coupled to Superior WH64 compressors. They also serve the power generation market by either supplying Waukesha's Enginator engine-generator packages or designing and manufacturing customized power generation systems to meet specific customer needs. One example is the four Model VHP7100GSI Enginators that supply 4.4 MW of power to the Canoas campus of the Lutheran University of Brazil near Porto Alegre.

More often than not, Waukesha engines are operating in can't-fail situations where even a brief shut down hits the customer hard in the pocketbook. It falls to the distributor to respond quickly and solve the problem. As Waukesha-Pearce vice-president Robert Lyde said recently, "Our job is to maintain maximum uptime, which requires a 24/7 mentality. Our response is as vigorous at midnight on a holiday as it is at 8:00 am on a Monday."

The distributor relies on the manufacturer to, first of all, build a quality product the sales force is proud to sell, then to back up the sale with prompt parts delivery and service training. As engines have grown more sophisticated,

3

4

4. The Landre & Glindermann exhibit at the Utrecht Trade Fair, 1965. Through L & G, Waukesha engines played a role in Europe's recovery from World War II.

5

5. The headquarters of Stemac, Waukesha's largest South American distributor, in Porto Alegre, Brazil.

6

6. Louis Pearce III, president of Waukesha-Pearce, Inc., and Robert Lyde, vice-president of operations 2004. Pearce heads up the family company founded by his grandfather in Texas. Waukesha Engine's oldest distributor, Waukesha-Pearce helped establish the Motor Company in the American and Latin American energy markets in the 1920s.

training has become more important. Today's technicians are just as handy with diagnostic software as they are with socket wrenches—perhaps even more so—and well they should be. PAMCO Enerflex, Waukesha's sole Canadian distributor has two-hundred-fifty factory-certified technicians and nearly two-dozen service locations in the gas fields of Alberta, Saskatchewan and British Columbia. "Our experience is unequalled in the natural-gas engine-driven business," said PAMCO vice-president Bill Moore. That experience pays off for PAMCO and Waukesha.

Waukesha-Pearce, PAMCO and Stewart & Stevenson are currently Waukesha's largest distributors, along with Kraft Power Corporation in the American Northeast and Reagan Equipment in the states bordering the Gulf of Mexico. They have been joined over the years by nearly sixty other distributors all around the world, each with the same commitment to excellence in customer service.

The number will certainly grow as yet-untapped natural gas resources are developed and the demand for decentralized and cogeneration power systems accelerates. As Waukesha-Pearce's Robert Lyde said, "Our objective is to be out there where our customers' horsepower is operating." Together Waukesha and its distributors work to get that horsepower out wherever it is needed and keep it running."

THE GREAT RECESSION

At the start of the Dresser era, Waukesha's sales department identified four markets for its engines. The construction and industrial market was the largest, with White (Oliver) Farm Equipment and Clark Equipment the leading customers. The petroleum market, which included both oil and natural gas, was a close second. Waukesha served this worldwide market primarily through American oil patch distributors led by Southwest Industries and Waukesha-Pearce in Houston, plus PAMCO in Denver, Colorado. Growth in the development of Mexico's energy resources had made PEMEX, the national oil company, another leading client. Together, the major customers in the construction, industrial and petroleum/natural gas markets accounted for over sixty percent of sales.

Much higher in profile, but much smaller in percent of sales, was the power generation market. Engines to be coupled to their own generators were sold to Onan, North American, Kohler and other equipment manufacturers, while Enginators went through distributors ranging in size from top seller W.A. Kraft in Boston to Singapore's Avery Laurence, which worked in the as yet minor market area of Southeast Asia. The major power generation customers made up about ten percent of Waukesha's total sales. The marine market was the smallest and accounted for about six percent of sales. Waukesha reached the commercial marine market through its petroleum sector distributors while sales to governments, mainly the United States and Mexico, were made directly from division offices.

In all its markets, Waukesha's leading competitors were Caterpillar, Detroit Diesel and Cummins, with Caterpillar accounting for about thirty percent of the approximately $650 million ($2.4 billion in 2006) non-captive engine market. Waukesha's share was six-seven percent. There was plenty of room to grow.

Growth did occur in the last half of the 1970s and continued into the '80s, with Dresser's Energy Processing and Conversion Segment hitting the $900 million sales mark in fiscal 1981.

Just as Dresser Industries was reporting a record $4.6 billion in sales for 1981, the American economy slid into the recession that plagued Ronald Reagan's first year in the White House. Officially, the "Reagan recession" lasted about sixteen months, but it had been preceded and accompanied by a slowdown in construction and followed by a severe contraction in agriculture and energy. The construction equipment industry had been struggling with the negative effects of high interest rates for years. Now a decline in commodity

1984
Waukesha produces its first Sulzer engine, a Model 16VAT25D.

1985
The ATGL series and the first Model GL lean burn, low emission, gas engines are introduced.

1986
Waukesha breaks ground for a new Training Center.

prices, the continuing consolidation of agricultural operations, and the well-publicized demise of the "family farm," crippled the farm equipment market as well. Two key niches for Waukesha engines—construction and agriculture—were hurting.

In the wake of the Iranian political crisis of 1979-'81, crude oil prices rose to $40 a barrel and every would-be wildcatter in Texas, for one, was drilling for oil. When prices fell to $11 per barrel, the wildcatters were hung out to dry. By the mid-1980s, hundreds of marginal rigs were shut down and the unemployment rate in Texas approached ten percent. The story was the same in Oklahoma, Louisiana, California and other oil-producing regions.

Sales of Waukesha engines plummeted. On one day in 1982, Chet Lee walked into the sales department and found no orders booked. "Do you mean to say," he asked, "that with all our distributors, all our contacts, no one in the world wants to buy a Waukesha engine today?" The answer he was forced to accept was, yes, no one wanted to buy a Waukesha Engine that day and on quite a few others. As a result, sales volume sank and by mid-1983, the number of employees on St. Paul Avenue fell below six hundred—lower than in any year since 1917, including the early 1930s. Employment numbers would rise again to peak at 1,200 in 1986, before falling below 1,000 by the mid-1990s and staying there.

The various recessions of the 1980s accelerated the pace of many ongoing trends at Waukesha, starting with a greater focus on fewer markets. The construction and farm equipment niches would go the way of the railroad and automotive niches. Just as Waukesha was no longer making Enginators for the Rock Island Line, which was itself no longer in business, or truck engines for the Four Wheel Drive Company, also no longer operating, Waukesha would soon stop building motors for White tractors and Clark fork-lifts.

Foreign competition was a major factor in the construction and agricultural markets. In the mid-1980s, an agent for one equipment manufacturer—and longtime customer—asked Waukesha to make an inexpensive "throwaway engine" to compete with the imports flooding the American market. Ben Stuart, who was division president at the time, refused, stating that Waukesha's reputation was based on quality. It did not build disposable engines. Stuart's seemingly prideful declaration was also a statement of company strategy. American manufacturers were finding it increasingly difficult to compete with foreign producers when a purchase depended solely on price. The Waukesha engine's reputation for enduring quality outweighed the cost and had served the company well for decades. It would continue to serve it well, but not in this market at this time. The equipment manufacturer stopped buy-

Ben R. Stuart

A twenty-five year veteran of Dresser Industries' domestic and international manufacturing operations, Ben Stuart served as president of Waukesha Engine from 1984 to 1987.

1987
Peter C. Trombley becomes President of Waukesha Engine.

1988
The VGF family of diesel and lean burn, low-emission, high-speed gas engines is introduced.

1989
The Clinton plant closes and Waukesha acquires Brons Industrie N.V. and its gas engine manufacturing plant in Appingedam, Netherlands.

Training Center

Product Training Manager Charles Lawton, in white shirt, (above), conducts a hands-on training session, 2006.

Waukesha built its Product Training Center on St. Paul Avenue in 1986. The 16,000-square-foot building houses classrooms that can be adapted to suit the many different courses taught and a shop area with work stations, tools, engines and generators for hands-on training. It is the home base for a Waukesha's worldwide natural gas engine training program.

ing Waukeshas and switched to throwaways, thereby hastening Waukesha's exit from the construction and agricultural equipment markets.

The marine market was also about to run aground, even though Waukesha's big engine lines were well-suited to shipboard applications. The loss of the contract to build an improved version of the Model L1616 non-magnetic engine for Navy minesweepers in 1986 was like a lightning bolt briefly but effectively illuminating the path to the future. In time, Waukeshas would go to sea only onboard offshore oil drilling or gas compression rigs.

While some markets might be closing, others were opening. Waukesha motors had been sold overseas since the 1920s so international marketing was not a new game for the sales staff. In the 1980s, however, developing international markets became more important than ever. If lightning struck here, the bolt came from a 1983 contract to sell twenty-five CFR engines to the People's Republic of China. The Communist regime was just initiating the non-Communist development policies that would make China the world's fastest-growing economy in twenty years. It was only logical that they would start by buying the testing instrument that would insure the quality of the gasoline that would fuel much of that growth. It was equally logical that Waukesha should look to expand its international markets well beyond twenty-five CFRs to Beijing.

By 1984, the days of the Waukesha diesel engine were numbered. Not long after he took over as president that year, Ben Stuart told reporters that

Waukesha's strength was in gas engines, not diesels. "Fortunately, we're not dependent on the diesel market for our growth," he said. "We have a unique niche in the gas market, where competition isn't so fierce. Waukesha Engine is the leader in gas engines." In the three years he served as president in Waukesha, Stuart would help make his projection a reality.

Another sign of future reality was the introduction in 1985 of the "lean-burn," low-emission GL models to the VHP family in order to meet tighter state and federal emission standards. The 650-horsepower, six cylinder Model 2895GL, the 750-horsepower 3521GL, the 1,300-horsepower, V-12 Model 5790GL and the 1,700-horsepower, V-12 Model 7042GL, accelerated the move to ever more fuel-efficient, environmentally-friendly engineering that would direct engine development for the rest of the twentieth century and into the twenty-first.

The final trend of the future that accelerated in the 1980s can be stated quite simply. Bigger is better. And it started with the 1984 Model AT25.

Lean-Burn

A VHP 7042GL on its way to New Mexico.

Night scene in the San Juan Basin of New Mexico, 1990s, where over two hundred VHP Model 7042GLs and other "lean-burn" Waukeshas power gas compression units. Waukesha introduced its first lean-burn, low-emission VHP models in 1985.

VGF Family

A VGF H24GL at the Sangre de Cristo Water Utility, New Mexico. The utility uses four of the engines to supply eleven million gallons of water a day to the city of Santa Fe. Waukesha acquired licenses from Guascor, the Spanish diesel engine builder, to *Waukesha-ize* their engines to burn natural gas and introduced them as the VGF family.

BIGGER IS BETTER

In 1980, Waukesha's most powerful VHP engines had yet to exceed the 2,000-horsepower mark. It was obvious to Chet Lee and his executive vice-president, Pete Trombley, that market demand would compel Waukesha to build engines twice the size of the largest VHP—and soon. They decided that they could not wait the estimated seven-ten years it would take for Waukesha to develop its own 3,000-4,000 horsepower engine. Instead, Trombley went on a global shopping trip, not just for a big engine, but for an engine that Waukesha's engineers could readily "gasify." They found it at Sulzer Ltd., in Switzerland.

Founded in the 1830s to manufacture industrial machinery, Sulzer was, in the 1980s, best known for its mammoth marine diesel power plants. At the lower end of its range was the AT25 Series which included six and eight-cylinder, in-line models as well as V-12 and V-16 engines that burned either light distillate or heavy diesel oil. In horsepower, they ranged from the 1,140-horsepower Model 6LAT25H to the 3,360-horsepower Model 16VAT25D. Waukesha purchased licenses to manufacture the AT25 and advertised its first Sulzer, a Model 16VAT25D in 1984.

Waukesha's first Sulzer engine may have been a diesel, but the intention was to turn out a gas burning model as soon as possible. By 1988, the entire AT25 family was available in both gas and diesel fueled versions, with the emphasis on the gas-fueled line. The push to gas engines was accelerated when

1990
Production underway on improved Model L1616 engines to equip minesweepers for the Royal Saudi Naval Forces.

1991
A Model L36GL engine powering a 520kW generator is installed at the plant to become Waukesha's first cogeneration unit.

1992
Waukesha Natural Gas Engines are Turning the World Around is the slogan appearing in print media and on division semi-trailers on the road in the United States.

Trombley went back to the European technological well to acquire rights to the Gauscor engine. A Spanish engine manufacturer, Gauscor produced a family of diesels in the 160-1,175 horsepower range that neatly bracketed the lower end of the VHP family, just as Sulzer's AT engines bracketed the top. As with the Sulzers, Waukesha purchased the Gauscors in order to modify them for gas. They ultimately constituted Waukesha's VGF family. By the end of the 1980s, the engine line was filled out with imported diesel technology Waukesha-ized to burn gas.

The trade was not all one way. The European market for gas-fired electrical generating units was growing and the continental nations were unifying their economies. To cite just one of many examples, in 1988 Waukesha shipped four Model L5108G VHP gas engines to power a cogeneration plant at a corporate computer center in Essen, Germany. To more readily participate in the new European Common Market, Waukesha purchased a manufacturing plant in the Dutch city of Appingedam in 1989 and began production of VHP and VGF engines there. At the same time, Waukesha completed the phasing out of its Climax engine plant in Clinton, Iowa. The Clinton plant had not been included in the 1970s renovation plan and it was only a matter of time before operations there were shifted to the more modern and efficient main shop.

The terms modern and efficient had come to mean more than just up-to-date manufacturing. As the cost of fuel and the scope of environmental regulation increased, it was only good business sense to develop engines that used less fuel and released fewer contaminants. The "lean-combustion" GL-engines

Water Power

The water pumping plant in the Sangre de Cristo mountains near Santa Fe.

1993
Volume One of the *Quality Digest,* Dedicated to the Total Quality Improvement Process, is published.

1994
G. Phillip Tevis is named president of Waukesha Engine and Appingedam becomes Waukesha's first plant to achieve ISO 9001 certification.

1995
Waukesha discontinues its diesel engine line to focus exclusively on gas-fueled engines.

Waukesha introduced were gas-fired VHP Models, ranging from the 650-horsepower Model 2895GL to the 2,063-horsepower Model 9390GL. They were turbo charged and intercooled like conventional Waukeshas but they also featured a patented prechamber that delivered a richer-air fuel mix to the combustion chamber. A thermostatic valve was added to the auxiliary cooling system to insure a constant air-fuel ratio, while high capacity turbochargers, low restriction intake manifolds and a larger carburetor fed the extra air necessary for the lean combustion fuel mix. As a result, a GL engine could churn out just as much horsepower as an equivalent conventional engine on nine percent less fuel.

In Europe

Waukesha's European manufacturing plant in the city of Appingedam, the Netherlands.

In part because they burned less fuel, but also because combustion was consistent and complete, the GL engines emitted dramatically smaller volumes of nitrous oxide, carbon monoxide and hydrocarbons. Since the majority of end users ran their units continuously, reduced fuel costs and lower emissions were a double-barreled bonus.

Harkening back to the Hesselman "all-fuel engine" Waukesha perfected its fuel systems so its gas engines could run not only on high-quality natural gas but on gas of inconsistent quality as well. This advance made it possible for gas compression units in remote locations to burn the same wellhead gas they were compressing. It also meant that the newest Waukesha power systems, whose predecessors had been running on digester gas at sewage treatment plants for years, could be installed at landfills to generate electricity by burning methane emitted by decomposing waste.

Concerns about cost, efficiency and emissions were key motivators behind the decision to halt the production of diesel engines in 1995. Waukesha had come a long way with diesels since the first automotive model was placed on display at the annual stockholders meeting in 1929. Now it was time to face forward once again.

Model L5774LT

Introduced in 2001, the L5774LT was the first VHP lean burn engine designed specifically for gas compression.

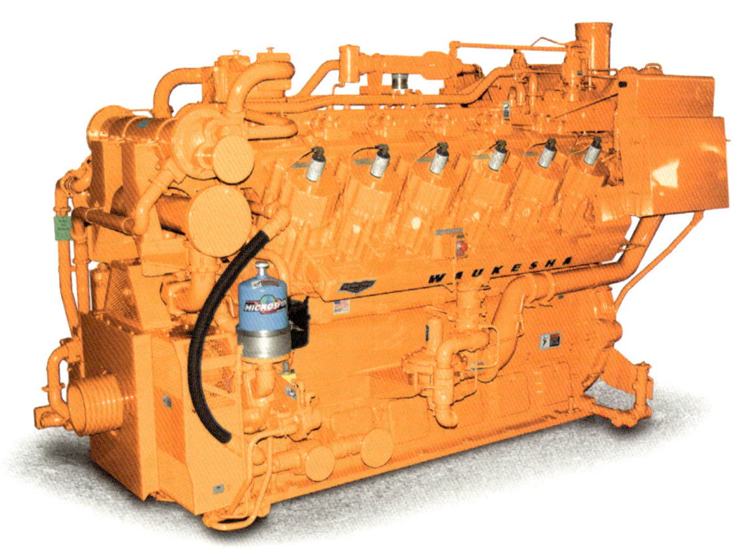

The year 1995 was a landmark as the end of the diesel era, but the previous year was memorable for the untimely death of fifty-four year old Pete Trombley. He had been named president of the division after Ben Stuart moved up the ladder at Dresser in 1987. As Chet Lee's executive vice-president and as president himself, Trombley had played a leading role in the reshaping of Waukesha Engine that began with the plant renovation in 1977, encompassed the start of the VGF and AT engine lines, and led to ISO 9001 Certification in 1995 and '96.

Trombley had been an early and strong advocate of the 1980's management philosophy that emphasized quality in every facet of operations. He instituted the Total Quality Management program in 1991, the Academy Award for Quality in 1993, and the first "Quality Day" held at the Milwaukee Zoo in 1994. Its theme was "Don't be a Dinosaur!", another way of saying that those who do not adapt face extinction. Trombley also initiated Waukesha's entry and eventual certification in the ISO 9001 program. In an article he wrote for the division's "Quality Digest" newsletter shortly before he died, Trombley stated that the purpose of the ISO 9001 program was to "establish a new quality system at Waukesha which puts our work force on a level with other world class manufacturers...

"The issue, by the way, is hardly academic. Some of our major customers are now asking tough questions about our processes, and how we go about taking care of business. They are comparison-shopping.... they are asking: How do you insure the integrity of your delivery promises? And how do you insure that spares will be available at the job site when the engine goes down or needs service?

"As we finish this important task, we must put emphasis on job training, so that we can get top quality from our shop and office employees the first time and every time...The real world of the customer-driven '90s is here and it is spelled ISO 9001."

Waukesha Power Systems achieved ISO 9001 certification in 1995 and the St. Paul Avenue plant in 1996.

Peter C. Trombley

"Pete"Trombley was president of Waukesha Engine from 1987 to 1994.

G. Phillip Tevis

G. Phillip Tevis served as president of Waukesha Engine from 1994 to 1996. He accepted the ISO 9001 Quality Assurance Certificate from Joe Marchese of Lloyd's Register in 1996.

Waukeshas at Work
1980s-1990s

1. In 1990, the Unocal Gas Storage Facility, at Aitken Creek, British Columbia, became the northernmost site Waukesha engines had yet been installed. The seven Model 12V-ATGL lean combustion engines successfully endured extremes of heat, cold and damp. Waukesha applied knowledge gained at Aitken Creek to refine the ATGL line to perform even better in extreme weather and still meet Canada's strict emission standards.

2. Two Waukesha VHP 7042GL units supply back-up power for the satellite tracking station in a remote location at Vandenberg Air Force Base in California. To insure reliability, the engines are designed to run on natural or propane gas. If natural gas service is interrupted, they can turn themselves off, switch to LP-gas stored on site, and power up again--all in a matter of minutes.

3. VHP engines drive aeration blowers at the Coldwater Creek treatment plant at St. Louis, Missouri, 1980s.

4. This VGF L36GLD is one of the one-hundred-plus Waukeshas at work for the EDON energy company in cogeneration applications throughout northern Europe since the late 1980s. They are part of a cogeneration system that supplies nearly one-third of the electricity in the Netherlands.

5. The Orange County, California, Sanitation District has twenty-one VHP L5790GU engines to drive pumps that move millions of gallons of sewage, 1980s.

1

2

2

3

4

4

5

Zoo Power

Two VHP 7100GL Enginators (below) provide back-up power for critical life support systems at Chicago's Brookfield Zoo. In addition to protecting the animals, the Waukeshas operate during peak demand periods and reduce the Zoo's utility bill. (Courtesy, Brookfield Zoo)

William O'Connor

William O'Connor was named president of Waukesha Engine in 1996 and served until 2003.

A VHP L5794LT at the Burlington Resources Canada Energy compression station near Edson Alberta (below), where the unit met the manager's goal of "99% run time."

THE NEW DRESSER

In 1997, Dresser Industries recorded $7.4 billion in revenue, about $1 billion more than in 1996 and $2 billion more than in 1995. The Power Systems Segment, which included Waukesha Engine and Dresser Roots, a manufacturer of centrifugal blowers and compressors, brought in $318 million. It was an increase over the previous years and, in fact, a record earnings year for the Waukesha Division, but nothing like the spectacular performance of Dresser as a whole. Dresser's neighbor in Dallas and occasional competitor in the energy service sector, the Halliburton Company, was also doing very well. It recorded revenues of $8.8 billion in 1997, about $1.5 billion more than in 1996.

In February 1998, Halliburton announced that it would acquire Dresser for about $7.7 billion in stock valued at $44 per share. *The Wall Street Journal* said the merger would "create a giant with $16.3 billion in revenue, 100,000 employees and a backlog of $13 billion in business contracts." In a news release Halliburton chairman and CEO, former Defense Secretary Richard Cheney, said "Halliburton and Dresser are an outstanding business and cultural fit. This is a win-win combination." Cheney became CEO of the new company, which would retain the Halliburton name, while Dresser's chairman and CEO, William E. Bradford, became chairman of the board.

Observers expected the new Halliburton, now the world's largest energy services company, to be a powerhouse, but it got off to a discouraging start. In 1998, the price of oil fell to $10 a barrel, the retail price of gasoline dipped below one dollar, and Halliburton recorded an $80 million loss. Waukesha's revenues declined, although not as severely as either of its parent companies.

William O'Connor continued to serve as president of the division. He had accepted the post in 1996, after heading up Dresser's Mono Pump Division in England for a number of years. His immediate predecessor, G. Phillip Tevis, who had succeeded Pete Trombley, had been named Dresser's vice president of operations.

Waukesha's Halliburton period was brief but eventful. As always, new designs, new technology and new engines came on line. The 1998 V16, 4,500 horsepower, Model AT27GL brought lean-burn design to Waukesha's most powerful engine. Also new were the Series Four models that expanded the VHP family. The 840-horsepower Model

1996
William O'Connor is named president of Waukesha Engine. The main plant achieves ISO 9001 certification and the first Waukesha website goes on line.

1997
Dresser Industries records $7.4 billion in sales, with help from a record-shattering performance from Waukesha.

1998
Dresser Industries merges with Halliburton to form a new Halliburton Company. Waukesha Engine is included as part of the Dresser Equipment Group.

3524, 1380 horsepower Model 5794 and the 1680 horsepower Model 7044 featured a new, stronger crankshaft and power cylinder plus a "state of the art" cylinder head that increased the engines' power and durability.

Waukesha was also making great strides in the field of engine controls, where computerized technology was bringing about profound changes. As far back as the 1980s, division engineers had developed Custom Catalyst Controls to "stand guard round the clock" on catalytic converters while reducing fuel waste and exhaust emissions. Waukesha also introduced its computerized Air/Fuel Ratio Control which could be programmed to run an engine "lean" on five-ten percent less fuel. They were a great leap forward from the Engomatic controls of the 1960s and a step on the way to the ESM control unit introduced in the late 1990s.

The ESM was designed to act like an engine-mounted technician that not only monitored the engine for optimum performance but could also report on it via an automatic communication system. It tracked engine speed, pressure, temperature and functions. It controlled the ignition timing and speed governor as well as stopping and starting.

One of the first ESM units placed in the field was installed on a VHP L7044GSI in operation at El Paso Energy's White Oak Creek gas blending and compression station in west-central Alabama in 2001. One of eleven Waukeshas at White Oak Creek, it powered a Dresser-Rand VIP four-stage compressor to pump low pressure methane coal field gas for pipeline transmission. The L7044GSI had been on the job for only six-months but El Paso agreed to the ESM field test. As in other gas pumping or compression applications, the engines at White Oak Creek are required to run 24/7, except for routine monthly maintenance.

The ESM couldn't eliminate the maintenance shutdowns, but it did all but bring an end to shutdowns for engine trouble. After a year with the ESM

ESM Engine Control

One of the first ESM engine control units was installed on a VHP L7044GSI at El Paso Energy's White Creek blending and compression station in 2001.

DRESSER Waukesha

1999
A $30 million manufacturing upgrade is completed, with $19 million spent on new computerized machine tools.

2000
Waukesha introduces the Model L5794LT, its first open-chamber, lean burn, VHP engine.

2001
U.S. Department of Energy grant starts development of APG engine line.
First Reserve Corporation and Odyssey Investments Partners, LLC, acquire the Dresser Equipment Group from Halliburton and organize the new Dresser, Inc.

Patrick M. Murray

Patrick Murray assembled the investment group that purchased Waukesha Engine and other Halliburton businesses to form Dresser, Inc. in 2001.

installed, El Paso's lead technician, Ray Settle, reported, "No problems whatsoever. In fact, other than routine preventive maintenance the engine has not been off-line all year long due to an engine failure. With the ESM we've enjoyed steady run times around 98%." By 2004, ESM controls were standard equipment on all VHP Series Four engines.

Halliburton's most famous chief executive resigned from the company in 2000 to join the successful presidential campaign of Texas Governor George Bush. In January 2001, Bush entered the White House and Richard Cheney moved into the vice-president's office. Back in Dallas, Patrick Murray, president of Halliburton's Dresser Equipment Group, was also working for a change in office. He assembled an investment group of Dresser executives and venture capitalist firms that offered to buy the Equipment Group from Halliburton for approximately $1.6 billion. Included in the deal were Dresser Valve Division, the Instrument-DMD-Roots Division, the Wayne Division and Waukesha Engine. They were all part of the old Dresser and solid businesses but not as strategically vital to Halliburton as, for example, Dresser's engineering, oil drilling and construction divisions.

The investment group saw a set of businesses that meshed well. "We were less concerned whether there was direct synergy between them and more concerned about whether in the broadest terms the market was similar," said Murray. All of the Dresser companies were in the energy business and all driven by the price of oil and gas. "We had a group of businesses that, if you draw the value chain, you've got upstream—drilling and production—you've got midstream—refining, power generation, gas processing—and you've got downstream, which is marketing." Together they possessed a "natural hedge" to correct temporary imbalances. "If, for example, Waukesha was doing particularly well—and 2001 was a really good year there—some of our other businesses were not doing so well, [the hedge] allowed us to balance the two." This is a textbook model of how a conglomerate is supposed to work and how the company Murray and his partners founded did operate.

With approximately $2 billion in revenue, the new company was large enough to obtain financing, fund a capital investment plan and offer greater protection to investors than a green start-up. Murray and his partners also purchased rights to the Dresser trademark, which was well known in the energy field and supplied a measure of brand equity. On April 15, 2001 Halliburton's Equipment Group became the new Dresser Incorporated, with Patrick Murray as president and chief executive officer.

"Enjoy the next several years," Murray told company newsletter readers that spring. "We've all earned this opportunity to make Dresser into the type of world-class organization that we've always wanted it to be."

World Trade Center Survivors

When it was designed in the early1960s, the World Trade Center in New York City was the world's largest commercial building complex. Seven buildings arrayed on sixteen acres would provide ten million square feet of office space for a work force of fifty thousand. The 110-story twin towers, One World Trade Center and Two World Trade Center, would soar 1,368 and 1,362 feet high. When completed in the early 1970s, they were the tallest buildings in the world.

No matter how reliable the conventional power system, a complex as large and with towers as tall as the World Trade Center required stand-by electrical power. When fully occupied, WTC would consume up to eighty megawatts of electricity, more than any stand-by system could reasonably supply. Therefore, to design a stand-by system, the Center's engineers identified core demands for power. They included the alarm system, lighting for the lobby, stairwells and exit signs, and six elevators in each tower to evacuate the upper floors. Preliminary estimates said a three thousand kilowatt system would do the job.

When asking for bids, however, WTC's builders did not limit themselves to the estimated specifications. They wanted the bidding to be as competitive as possible to obtain the best power system they could find. Working with the electrical contractor Hatzel & Buehler and Waukesha's northeastern United States distributor, W.A. Kraft Corporation, Waukesha Power Systems assembled a package of five Enginator units and was awarded the contract. The engines were VHP Models L5792DSI, twelve-cylinder, turbo charged, inter cooled diesels. The generators were rated at 1,000 kw continuous, 1100 kw standby. A sixth unit was later added and all were uprated to 1,200 kw standby.

They were installed in the sixth sub-basement of One WTC, seventy feet below ground level. A railroad siding ran down to the basement which was designed with doorways large enough to remove the engines in the unlikely event they needed off-site servicing. Cooling water came directly from the nearby Hudson River and exhaust was vented to the surface via a six-story tall exhaust pipe.

When terrorists first attacked WTC in 1993, they set off a truckload of high explosives in the parking garage four stories above the Enginators. The blast severed much of the wiring between the units and the building above, ruptured the cooling water lines from the river and knocked out the basement drainage pumps. Nonetheless, personnel from WTC and W.A. Kraft kept the units running for about three hours, until the floodwaters rose to the crankshaft centerline and the engines had to be shut down. The system was designed to cope with blackouts, fires and other emergencies occurring in the buildings, not a truck bomb detonating in the vicinity of the Enginators themselves.

No one anticipated the horrible force of the second terrorist attack on September 11, 2001. Since some lights remained on as the towers collapsed, it appears that the Enginators switched on and ran until One WTC came down on top of them. Had the buildings remained standing, the system should have been able to meet the demand for emergency power.

Of the six Enginators beneath One WTC, three were damaged beyond repair. One of the three was brought back to Waukesha and is part of the Waukesha Engine Historical Society's collection. Three were salvageable and have been rebuilt, rugged veterans of the worst act of terrorism inflicted on the United States.

One of the surviving Model L5792DSI Enginators on exhibit at the Veterans Memorial, Milwaukee, in the summer of 2002.

TBG Across Brazil

Three VHP L5790GSI engines are among the forty-two Waukeshas moving natural gas two thousand miles from wells in Bolivia to populous coastal cities of Brazil. The gas-fueled engines help make it possible for the South American nations to export oil and use their gas resources to meet domestic needs.

IN A NEW CENTURY

The "next several years" started pleasantly, then plunged into the grim nightmare of the September 11 terrorist attacks. The fact that Waukesha Enginators were part of the World Trade Center complex only brought the tragedy closer to home. When one of the units returned to Waukesha in 2002, it was a time for muted celebration and reflection.

The economic downturn that followed September 11 came in the wake of the energy crisis in California and the end of the 1990s bull stock market. It was followed by a decline in natural gas prices compounded by the Enron scandal and one of the warmest winters recorded in North America. Waukesha's revenues declined from $277 million in 2001 to $217 million in 2002, contributing to a lay off of one-hundred-thirty workers. Waukesha's financial performance was tied to the price of natural gas. When gas was up or even when gas was forecast to go up, so did engine orders. Since gas prices had bottomed out in the summer of 2002, William O'Connor cautiously predicted an upturn starting in the second half of 2002 and continuing into 2003 and 2004. As time would tell, he was correct.

Demand for back-up power units increased, stimulated by the terrorist threat, by the California crisis and by the power failure that struck six states

2002
The Model L5792DSI engine that survived the destruction of the World Trade Center comes home to Waukesha.

2003
Labor-management contract dispute is settled after an eighty-five day Machinist's union strike.

2004
Thomas J. Laird becomes president of Waukesha Engine and Waukesha purchases technology rights to the Wartsila W200D and W220SGD engines.

and one Canadian province in the summer of 2003. During that blackout, Waukesha Enginator units kicked on to produce 100 MW of electricity for businesses, hospitals, municipal services and crisis control centers from Michigan to New York. The city of Chicago was not affected by the blackout but it was prepared. Chicago had already built a secure 911-emergency communication and crisis command station with two VHP 7100GSI Enginators in place. They insured that the 911-system would not go down when needed and other city emergency services could function in a crisis.

TBG Pipeline

The compression station at Florianoplis, Brazil, on the TBG pipeline.

Continued overseas development also aided Waukesha's recovery. In 2002, the TBG Pipeline, which carried natural gas from fields in Bolivia to Rio de Janeiro, Sao Paulo, Porto Alegre and other population centers on the coast of Brazil, was completed. Forty-two Waukesha engines ran gas compressors and electrical generators along the 2,000 mile route from the Andean highlands to the sea. With the largest, fastest-growing economy in South America, Brazil was becoming a major market for Waukesha in both the gas compression and electrical generation fields. The Brazilian state energy company, Petrobras, opened a technical training center in the economically depressed but natural-gas-rich northeast region. Located in the coastal city of Natal, the Gas Technology Center could train as many as five thousand students who could make use of the Model VGF H24GL engine running the Center's powerplant as a teaching aid.

It was a logical step for Waukesha to open its own training center in Brazil. In November, 2003, Waukesha launched the Latin American Product Training Center at Porto Alegre in cooperation with Stemac, its authorized distributor for Brazil. Waukesha had over 3,600 engines operating in Latin America, and the Training Center would serve distributors and end users from Mexico in the north to Argentina in the south.

Building on its South American experience and as a consequence of restrictions placed on travel to the United States after the September 11 attacks, Waukesha opened a training center to serve the expanding Asian market at Phitsanulok, Thailand in 2006. It is located on the premises of Thai distributor, Souer Company, Ltd., and trains technicians from a sizable wedge of the planet stretching from Japan south to Australia and west to Pakistan. South Asia—Bangladesh, India and Pakistan, where approximately one thousand Waukesha units were already in place—offered the greatest promise. In 2002, four VHP-powered Enginators started operating at the Arif Knitspin mill

2005
The Model APG1000 goes into field testing.

2006
The APG 1000 is released for sale. Waukesha celebrates its centennial.

in Dhaka, the capital of Bangladesh. The units supplied four megawatts of power to run machines that spun raw cotton into thread. It was another 24/7 application that required reliable, uninterrupted power, which placed it squarely in Waukesha's bailiwick. Not long after the first four VHP units were running, Arif Knitspin ordered two more.

In the midst of the recovery, Waukesha encountered its most serious labor crisis since 1939. In May 2003, about four-hundred-sixty members of Local 1377 of the International Association of Machinists and Aerospace Workers voted to strike. They had rejected a company proposal introducing a two-tiered pay scale, additional flexibility in short-term layoffs and a retiree contribution for health insurance. The proposal reflected the environment in which all American manufacturers were now operating. They had to face competition from low-wage foreign companies. They had to push constantly for increased productivity and quality, which required greater flexibility in all operations. In addition, health insurance costs for employees and retirees had been rising and would continue to rise with no end in sight. The company was looking to the union to share some of the costs and partner in agreeing to new work rules that would enable them to be more competitive.

The union employees went on strike, but 2003 was not 1939. Non-union staff and temporary workers crossed the picket line and the plant did not shut down. Talks continued, with the union eventually accepting the two-tiered wage plan and a change in layoff rules, while the company accepted a reduction in the amount it would require retirees to pay for health insurance. The strike had lasted eighty-five days and, like a summer thunderstorm, had briefly cleared the atmosphere, but the issues it had raised would not blow away. More than a year before the 2003 contract was set to expire, the union asked to begin negotiations on the next agreement. In the fall of 2005, the company and Local 1377 worked out an unprecedented five-year contract with new provisions to address market competitiveness and skilled worker shortages. It also included an annual wage increase and preserved benefits for retirees. Union employees would continue to make substantial contributions to their medical coverage. If Waukesha was going to thrive in the 21st Century, everyone had to be on board.

Everyone also had to be on board because, as Waukesha approached its centennial anniversary, exciting things were happening. In 2001, the company had qualified for a Department of Energy grant as part of the Advanced Reciprocating Engine Systems (ARES) program. The goal of ARES was to support the development of high efficiency, low emission engines for power generation. Waukesha's engineers went to work, designed and completed test models of the new Advanced Power Generation (APG) engine early in 2005. The new APG engine was a turbocharged, intercooled V-16, with a lean-combustion gas fuel system. It was designed to run 1,000 kw generators, which occupied a significant market niche. In the words of Jim Drees, vice president of product development, the APG1000 would "go after the very large market segment of the power generation market that demands high efficiency, low emissions and lower operating costs simultaneously."

Developing South Asia

Four VHP L5794GSI and L7044GSI Enginators supply the power to keep the mills spinning cotton yarn into cloth at the Arif Knitspin mill in Dhaka, Bangladesh. The south Asian region, including Bangladesh, Pakistan and India, is one of the most promising markets for Waukesha power units.

Field testing began almost immediately, first at Waukesha, where an APG1000 was installed in parallel with the local grid to provide electrical power to the testing labs and the manufacturing plant. Two additional units were installed in real life applications at factories in Ohio. Production models of the APG1000 were scheduled to debut by the end of 2006. "We met our thermal efficiency requirements for the program's first phase, achieving over forty-two percent thermal efficiency," said Jim Drees in late 2005. "In Phase Two we'll be focused on attaining a goal of forty-seven percent efficiency."

While improved efficiency was the goal of ARES, Waukesha expanded the APG program to develop more powerful 2,000 kw and 3,000 kw engines. To produce these engines, Waukesha approached Sulzer's corporate successor, Wartsila, Ltd. and purchased technology rights to the Wartsila W200D and W220SGD engines. They were designed to run generators ranging from 1,800 to 3,200 kw, well inside the parameters of the expanded APG program. In 2006 Waukesha started work on modifying the Wartsila W220 design into a new Waukesha Model 18V220GL. Rated at 4,425 horsepower, it will be the most powerful engine the company has built. All of the Wartsila designs will be Waukesha-ized by the addition of state-of-the-art ESM control systems.

In order to build the 18V220GL and the other APG engines, Waukesha/Dresser would invest $3 million to improve manufacturing facilities at the St. Paul Avenue plant and $7 million to install new computer numeric controlled machine tools. Merely to handle the twenty-five-foot-long, ten-foot-high, forty-one-ton Model 18V220GL would require improvements to the loading dock and a new overhead crane powerful enough to hoist the behemoth. Additional capital investment is planned that will have an impact far beyond the manufacture of these big engines.

In its centennial year, Waukesha reflected briefly on a century of engine development, then moved on to the work of the future. From the Model A to the Model APG, the differences are phenomenal. Yet not everything has changed since 1906. Harry Horning and his partners set out to build a better engine than any they had seen. As evidenced by the APG, Waukesha continues to produce new and better engines today. Early on, the founders explored the bounds of available technology, adopted or adapted it as needed, or created it anew, just as today's engineers will Waukesha-ize the Wartsila W220 with their own patented controls. Horning and company also realized that, no matter how well conceived, engineered and built, their engines must compete in the market place. Today Waukesha focuses on its strongest markets—gas compression and power generation—and designs, builds and sells engines for those markets anywhere in the world.

For Harry Horning and those who followed his lead, it was not enough just to build motors. They had to build an engine company. They faced forward and, a century later, Waukesha Engine faces forward today.

APG 1000

Waukesha introduced its Advanced Power Generation 1,000 kw engine unit in 2005. The high efficiency, low emission APG 1000 was the first model in a new family of engines.

Model 18V220GL

In 2006, Waukesha acquired rights to the Swiss-designed Wartsila W220 diesel design in order to *Waukesha-ize* it for natural gas fuel. Rated at 4,425 horsepower to run 3,000 kw generators, it is the the most-powerful engine Waukesha has yet built and represents a step on the way to bigger and better engines in the future.

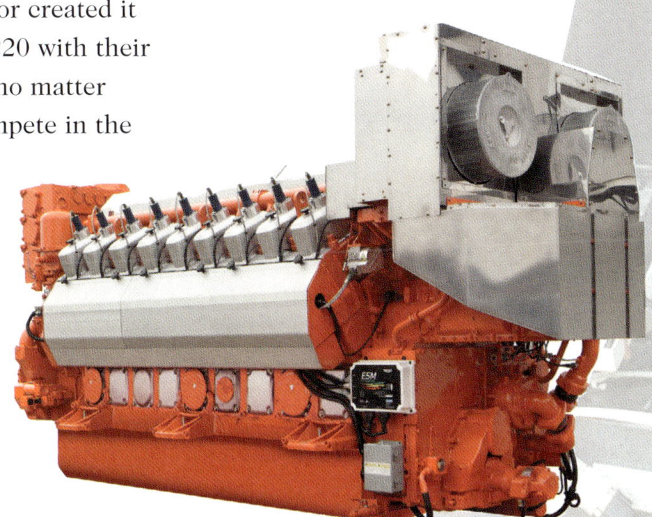

Global Delivery

A Waukesha VHP unit bound for gas compression duty in Poland is loaded onto a giant Antonov-120 air freight plane at the airport in Calgary, Alberta, Canada.

CHAPTER FIVE

FACING FORWARD

THE LIARD RIVER ARCS OUT OF THE MOUNTAINS IN CANADA'S YUKON TERRITORY, DIPS SOUTH INTO THE PROVINCE OF BRITISH COLUMBIA AND JOINS THE GREAT MACKENZIE RIVER TO FLOW NORTH TO THE ARCTIC SEA. AT A CLEARING IN THE boreal forest of the Liard Valley, Devon Energy operates a gas pumping and compression site called Kotalneelee. It is about two thousand miles north and three thousand miles west of Waukesha. Devon opened Kotalneelee when energy prices rose in 1979 and equipped the site with two VHP F2895-powered generator sets. When energy prices fell in the early 1980s, Devon shut down Kotalneelee.When prices rose again in the early 1990s, Devon reopened the site and was able to rely on natural pressure to push the gas up its 4,200-foot deep wells and into the processing and pipeline system that carried it to markets in western Canada and the United States.

By the end of the 1990s, Devon could no longer depend on natural pressure to move Kotalneelee gas and, working with Pamco, Waukesha's Canadian distributor, installed a compression unit driven by a Waukesha Series Four VHP L7044 engine. Kotalneelee can produce thirty-nine million cubic feet of gas a day. Running 24/7, the L7044 burns a fraction of that wellhead gas to compress the remainder and send it down the line.

Tens of thousands of miles south and east of Kotalneelee—and Waukesha too—the Magdalena River courses through the heart of the South American nation of Colombia. It also flows north, not to the Arctic, but to the Caribbean Sea. Colombia has not been known as an significant energy exporter, but as world demand has increased, so has development of reserves in ever more remote places, like Kotalneelee, and in places where production costs were once prohibitive, like Colombia. There are well over one hundred Waukesha engines pumping oil, compressing gas and supplying electricity in this developing nation.

Thomas J. Laird

Named president of Waukesha Engine in January 2004, Tom Laird presided at the inaugural of the APG1000 engine in 2005.

Like other South American energy producers, Colombia exports as much oil as it can while reserving its natural gas to produce that oil, fuel electrical power plants and meet other domestic energy needs. As a result, oil has surpassed coffee as Colombia's number one export and source of foreign currency.

Waukesha engines mix into this strategy like Wisconsin cream in Colombian coffee. In the uplands near the city of Neiva on the Magdalena, is the Campo Matachin oil drilling site. One Waukesha VHP F3521GSI and three VHP P9390G engines burn wellhead gas to generate power to pump twelve thousand barrels of oil a day and pipe it to production facilities a few miles away in the aptly-named village of Purification. Here three gas-fired VHP 7042GSI engines power equipment that removes water from the oil before sending it down the line for export. Surplus natural gas is also processed at Purification and sold in Colombian markets.

On the other side of the planet, just as far from Campo Matachin as it is from Kotalneelee, is the city of Poznan in western Poland. It is located on the Warta River, which flows north to the Baltic Sea, in what is becoming one of the most important natural gas transmission centers in Europe. Western Poland is an energy crossroads where gas lines from Russian Siberia in the east meet pipelines from the North Sea on the west. Once there, the gas is distributed throughout Europe, where it is the fuel of choice for power generation and industry.

The Waukesha engines operating in Poland are a mix of late-1990's VHPs and newer ATGLs. At the Zuchlow gas compression station near Poznan, they process eighty thousand cubic meters of gas an hour from nearby wells for the Polish market. At other stations in the region VHP P9390GSIs compress gas for transmission across Europe and Russia.

Poland itself is one of the largest generators of electrical power in Europe. Nearly all Polish power plants are coal-fired and were built by the former Communist government with little attention paid to the environmental standards of the European Union. As the Polish economy is integrated into Europe's, its power system must measure up. To help meet stricter standards, Poland plans to expand its natural gas-fired electrical generating system. Through peak-shaving and co-generation, Polish energy companies hope to use natural gas to

North To The Arctic

The Kotalneelee gas compression site in Canada's Yukon Territory.

generate fifteen percent of the nation's electricity by 2010. As Poland develops an energy infrastructure comparable to that of western Europe, more Waukesha engines will come into play for both compression and generation.

An excellent model for the Polish generation system of the future can be found less than one hundred miles away from Waukesha at Northwest Community Hospital in the Chicago suburb of Arlington Heights, Illinois. In the mid-1990s, Northwest installed three Waukesha VHP 7100 Enginator units. It was not unusual for a hospital to install an Enginator package for backup power, but Northwest had more than emergencies in mind. Expecting the cost of electricity purchased from its local utility to rise, Northwest made a strategic decision to reduce its electricity bill by investing in natural-gas-fueled Waukeshas.

At Northwest the Enginators work the day shift, 9:00 am - 6:00 pm, when hospital demand peaks and when the local utility charges its highest rates. By shaving the amount it purchases from the utility, the hospital pares down its energy bill. In addition, heat from the engines is captured to cogenerate steam heat for other duties, such as sterilizing equipment. Through peak-shaving and cogeneration, and depending on the price of the gas fueling the Enginators, Northwest Hospital has saved up to $650,000 a year in energy costs.

The system worked so well that, when the hospital expanded its facilities in 2005, it also expanded its power plant by installing the newest Waukesha, an APG 1000. More efficient than the VHPs, the APG 1000 helps the hospital further reduce its energy bill. As a low-emission engine, the APG 1000 also helps Northwest Community Hospital meet its commitment not only to operate economically but also in an environmentally-sensitive manner.

Kotalneelee. Matachin. Poznan. Arlington Heights. Separated by many thousands of miles, they nonetheless indicate the direction Waukesha Engine will face in the future. The perspective is global from start to finish—from wellheads in places as remote as the Yukon to power plants in sprawling metropolitan areas as developed as northeastern Illinois. "It's a much more global world than it was twenty-five years ago," observes Thomas J. Laird. He was

On The Magdalena

A VHP F3521GSI generator set (above, left) supplies electricity to pump oil and gas at Campo Matachin in Colombia, South America.

Northwest Community Hospital in suburban Chicago (above) has used Waukesha VHP Enginators to supply uninterrupted power for vital services and to reduce energy costs through peak-shaving and cogeneration since the mid-1990s. The addition of a high-efficiency, low-emission APG1000 unit in 2005 supports the hospital's plans for an environmentally-friendly "green" expansion.

Reliable Power

Technicians at the Energy Works in Valencia, Venezuela, where nine VHP Enginators produce ten megawatts of electricity for Ford Motor Company to operate one the largest automotive assembly plants in South America.

named president of Waukesha Engine in January, 2004, after thirty-plus years in marketing, operations and management with General Electric and other manufacturers. A thoughtful student of management theory, Laird's goal is to insure that Waukesha's operating procedures are as up-to-date as its engineering.

No where in business is a global view point more important than in the energy sector. For Laird, "The beauty of Waukesha is that, number one, we're in natural gas. We're at the beginning of the process and we're at the end of the process. The whole energy pipeline."

Waukesha is at Kotalneelee and Matachin, two places at the "beginning of the process" whose differences extend well beyond the weather in January. The composition of the gas at Kotalneelee is not the same as at Matachin and Matachin's is not the same as that found in Poland, Kazakhstan or the Gulf of Mexico. What all these varieties of gas have in common is that a Waukesha engine can be designed to run on any and all of them. "Engines in the gas compression market are out there pulling field gas out of the ground," explains Laird. "Raw 'field gas' can have all kinds of contaminants and the quality can vary widely. That presents a real challenge for an engine manufacturer and some of our competitors don't do as well with field gas as we do. It's been a competitive advantage for us."

In September 2006, energy consumers were pleased to learn that a consortium of drilling companies, including Devon Energy, had discovered a reservoir of oil five miles beneath the bottom of the Gulf of Mexico that could double American reserves. Where there is oil, there is natural gas. It will take a few years to bring this huge field into production, but it is all but certain that when Devon and its partners are ready to pipe the gas, Waukesha will have engines designed to run on it.

As the histories of Kotalneelee and Matachin illustrate, the price of energy governs the market for natural gas and, of course, investment in equipment to bring the gas to market. "If you can tell me what the natural gas price will be in five to ten years," says Dresser, Inc. President Patrick Murray, "I can tell you how much Waukesha will grow." As energy prices rise, development of oil and gas reserves expands at both new and old fields. The future is in the market and, in 2006, the market at the start of the energy chain looks promising for Waukesha.

Waukesha is also well-positioned at the end of the chain. Northwest Community Hospital sets the standard, but other examples abound of Enginators installed to insure an uninterrupted source of power, reduce costs or utilize an otherwise wasted resource.

Valencia, a city of three million in Venezuela, is an automobile manufacturing center. In 1999, Ford opened a plant to assemble trucks, cars and vans that would require up to ten megawatts of electricity. Choosing not to rely on the local utility, Ford contracted with Energy Works, an American company, to build and operate a power station near its plant. Working with Lukiven, Waukesha's Venezuelan distributor, Energy Works acquired nine VHP7100GSI Enginators, each capable of generating 1.1 MW. At least three Enginators run constantly to meet the plant's base demand for power. When the Ford plant is running full bore, all nine Enginators kick into gear.

The power supply is not always reliable in Venezuela, and sometimes in the United States as well. In New York City power outages may be infrequent, but they happen often enough to convince the Home Depot chain to install Waukesha Enginators at seven big-box stores in the area. They keep the lights on, the ventilation fans blowing and the computers running no matter what bump trips up the grid.

It is not relevant in New York, but the fact that Waukesha engines can burn gas of varied quality can also be an asset for power generation. In southeastern Wisconsin, Superior Services, one of the largest waste management companies in the United States, operates the Glacier Ridge landfill. The site had been used as a municipal "dump" for years and was venting methane gas that had to be wastefully flared off. Not long after it took over the site in 1993, Superior installed two VHP5904LTD Enginators that turned the methane energy into two megawatts of electricity. The Waukeshas supply all the electricity for the landfill and generate a surplus that Superior sells to the local utility. The quality of landfill gas can fluctuate widely, yet the Glacier Ridge engines have performed as well as expected for over a decade.

Versatility in fuels is important but the two most powerful forces steering engine development today are efficiency and emissions. The market wants more of the first and less of the second. For Laird, "Higher efficiency means less fuel and more power. That's the market saying we're got to get a little bit better every day. Ten years ago we were in the thirty to thirty-five percent efficiency range. Today the APG is forty-two percent. We are developing forty-seven, forty-eight percent for five years from now. We don't think there's anything to slow that process down."

Laird sees efficiency as only one side of the coin. "There's another challenge on the other side. How do you make an engine with ultra-low emissions? That's another world wide market trend that started very slowly but it's going to really accelerate because of environmental concerns."

The customer concerned about emissions may be in China, with its huge coal-burning power plants and cities with the most heavily polluted air on earth, or in a place like Janesville, Wisconsin, where the city government has followed the lead of Superior Services at Glacier Ridge and installed a set of Waukesha engine-driven generators to burn methane from the municipal landfill to generate three megawatts of electricity. In the USA, China and elsewhere, demand for cleaner energy continues to grow. That bodes well for Waukesha.

They are still mechanical devices, but today's engines are controlled by electronic technology. Developing this technology will also be a large part of Waukesha's future. Today's ESM control units can alert the operator to signs of trouble. Tomorrow's controls will be more sensitive and specific. Laird describes how it might work: "We want the controller to electronically send a request to the maintenance department to come in and maintain cylinder Number 14 of engine Number 13. It's got eight hundred hours before failure.

Versatility

The Glacier Ridge landfill in Wisconsin, where a VHP generator set produces electrical power from formerly-wasted methane. Readily able to burn gas of varying quality, Waukesha engines are ideal for landfill applications.

"The pieces of the technology are available today. Now we've got to tie it all together so that we have an engine that talks. Think about it that way. The engine can talk to us, the supplier, and tell us what's working and what isn't. That's valuable information for future design. The value of engine monitoring is virtually unlimited in terms of Waukesha, our customer and our suppliers.

Patrick Murray pushes the scenario farther: "Let's engineer the failure out of the engine." Most gas compression and generator applications require more than one engine, not just to meet power needs, but also for back-up, and for an obvious reason. "The well is running 24/7. When that well is shut down, it's a lot of money lost." Murray proposes that "rather than paying for redundancy," end users have access to engines that are failsafe. "If you have a one-hundred percent certainty that one engine is going to run ninety-eight percent of the time, that's far better than having two or three of them stacked up in the event one or the other goes down."

A near one-hundred-percent-uptime, ultra-low-emissions, failsafe engine sounds like pie-in-the-sky, but a century ago, so did the notion that the gasoline motor would replace the steam engine and, a century earlier, that the steam engine would replace the draft horse. The history of Waukesha Engine is one of technological innovation. So is its future.

It is also a matter of people, starting with management. For Waukesha's top manager, "Leadership is important," Laird says. "As long as the leadership focuses on the realities of the market, the realities of the world we live in, and remembers that it is constantly changing, we have a good opportunity to be successful."

Teamwork is more important than ever. "The employees within the organization have to work as a team like they never have before. We have a much more competitive marketplace and we depend so much on employees, not just their manual efforts but how they think. We really need to empower employees and bring them onto the team at a whole different level."

People Power

As with its past, the future of Waukesha Engine is in the hands of its people.

After three decades of automation and computerization, a Waukesha Engine employee today is as productive as twenty or more shop workers of the 1970s. More than ever, that worker is part of the critical chain of procurement, engineering, manufacturing, marketing, sales and service that constitutes a successful enterprise in the 21st century.

"When an operator is engaged in his operation, we need to make sure the materials, the tools, the fixturing are right at his fingertips. We're focusing on the individual work stations and the individual operators... they have to have a higher level of skill." Laird may be referring to assembly workers here, but his concept extends to everyone at Waukesha Engine, from service desk to shop floor to executive office—and beyond. "One of our critical competencies will be training people in the whole supply chain." The future of Waukesha will be determined at the component manufacturer, at the home plant, and at training centers a continent away.

The conclusion is inevitable. "It's really about the people of Waukesha. The people of Waukesha have brought this business where it is. Its future is absolutely dependent on them."